THE **PRACTICAL**
ASTRONOMER

THE PRACTICAL ASTRONOMER

WILL GATER AND **ANTON VAMPLEW**
CONSULTANT **JACQUELINE MITTON**

Penguin Random House

THIRD EDITION

Senior Editor Peter Frances
Project Editor Tina Jindal
US Editor Kayla Dugger
US **Executive Editor** Lori Cates Hand
t **Editors** Rupanki Arora Kaushik, Francis Wong
Assistant Art Editor Arshti Narang
Editors Angeles Gavira Guerrero, Rohan Sinha
g **Art Editors** Sudakshina Basu, Michael Duffy
signers Rakesh Kumar, Umesh Singh Rawat
re-**Production Manager** Sunil Sharma
Production Editor Kavita Varma
Production Controller Meskerem Berhane
nior **Jacket Designer** Suhita Dharamjit
Design Development Manager Sophia MTT
ociate **Publishing Director** Liz Wheeler
Art Director Karen Self
Publishing Director Jonathan Metcalf

SECOND EDITION

Senior Editor Peter Frances
ditors Frankie Piscitelli, Miezan van Zyl
US Editor Kayla Dugger
Art Editor Mark Lloyd
naging **Editor** Angeles Gavira Guerrero
Managing Art Editor Michael Duffy
Jacket Designer Mark Cavanagh
e-**Production Producer** David Almond
Senior Producer Anna Vallarino
Associate Publisher Liz Wheeler
Art Director Karen Self
Publishing Director Jonathan Metcalf

Produced for Dorling Kindersley by

cobaltid

www.cobaltid.co.uk

Editors Marek Walisiewicz, Kati Dye,
se Abbott, Sarah Tomley, Robin Sampson
Art Editors Paul Reid, Lloyd Tilbury,
rren Bland, Claire Dale, Rebecca Johns

This American Edition, 2020
First American Edition, 2010
hed in the United States by DK Publishing
Broadway, Suite 801, New York, NY 10018

For the curious

www.dk.com

CONTENTS

INTRODUCTION

Humans have gazed up at the sparkling stars since time immemorial, trying to understand this immense, untouchable extension of our natural surroundings. In ancient times, the unknowable nature of space was cause for both wonder and fear. Eclipses were viewed with terror: why was the Sun slowly disappearing, the sky darkening, and the air turning colder? Even now, ancient superstitions remain in many cultures around the world regarding eclipses and comets. It seems that no matter how much science and technology demystifies space, the human consciousness continues to be fascinated by it. Our interests fall across many disciplines: as well as exploring the chemistry and physics of the stars and galaxies, we are looking for answers to more philosophical questions about the beginning of the

Universe and the possibility of life beyond our planet. And always we are aware of the sheer beauty of the Universe's celestial objects—many of the images taken by telescopes such as Hubble are breathtakingly artistic.

Whatever your reason for wanting to stargaze, your starting point is to find out what there is to see, when to look, and how to find it. Of course space does not always play a set-timetable game; a comet, for example, can appear unexpectedly in the sky, but there are a multitude of astronomers with telescopes trained on the skies every night who will quickly alert the world about the event via the Internet. This book will help you become a valuable part of that astronomical community. The first part of the book explains the kinds of objects you'll be looking at, such as planets, stars, and nebulae, and shows how to navigate around the night sky, treating it as a sphere. This is followed by practical considerations when buying and using telescopes, together with information on sketching and astrophotography to help you record your observations. The book then focuses on what you will be seeing, providing maps of the constellations, starhopping charts, and detailed information on the planets and stars of our own galaxy. Once you get started, you'll be surprised at how much you can find and understand in the night sky. Within just a short time you will be able to find planets, identify stars, track movements, find constellations, and even begin starhopping from one constellation to another. Good luck on your stargazing quest.

1

UNDERSTANDING THE UNIVERSE

The Universe encompasses everything that exists, from the tiniest of atoms to the largest of galaxies. In watching the night skies, astronomers can trace the signs of the Universe's beginning and learn about its evolution through studies of the stars, planets, nebulae, and galaxies.

THE UNIVERSE

The Universe is the name astronomers give to the vast space which contains everything we know, from great clusters of galaxies to the tiniest of particles. Professional astronomers have gathered clues about its evolution by capturing the light from galaxies many billions of light-years away.

EXPLORING THE UNIVERSE FROM EARTH
Telescopes such as the ground-based Gemini South telescope in Chile have enabled astronomers to peer into the very depths of the Universe to reveal its origins.

The Big Bang

The Universe is thought to have been born 13.8 billion years ago in a searing hot maelstrom known as the Big Bang. One strong piece of evidence for this energetic beginning comes in the form of a ubiquitous source of microwave radiation appearing across the whole sky, known as the Cosmic Microwave Background (CMB). As the Universe has stretched and expanded since the Big Bang, it has also stretched the wavelength of the radiation emitted when it was very young. So what was once higher-energy, shorter-wavelength radiation now appears as lower-energy, longer-wavelength microwaves. Tiny fluctuations in the density of matter early on in the Universe acted like seeds around which matter began to clump, setting the scene for the formation of vast clusters of galaxies. The fingerprints of these "seeds" can be seen stamped as minute variations in the Cosmic Microwave Background.

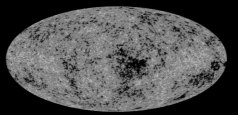

MICROWAVE CLUES
The Cosmic Microwave Background (CMB) radiation appears in all directions across the sky. The WMAP image maps it in incredible detail, showing the minute variations in its signal. The CMB was first discovered by Arno Penzias and Robert Wilson in the 1960s.

The first stars

The first stars are thought to have appeared around 180 million years after the Big Bang. They would have been truly gargantuan objects—possibly 100 to 1,000 times the size of our Sun, and made almost completely of hydrogen and helium. The stars we see today in our own galaxy and in nearby galaxies are totally different from this first generation of stars. This is because over billions of years, generations of stars created and dispersed new chemical elements as they lived and died, enriching the Universe with heavier elements. Carbon, oxygen, iron, and silicon were formed from nuclear fusion in the hot cores of the stars, and the very heaviest elements were forged by supernova explosions. As each star dies, it scatters new enriched star-forming material across space. Without these successive generations of stars, we simply would not be here today, as the ingredients to make us would not exist.

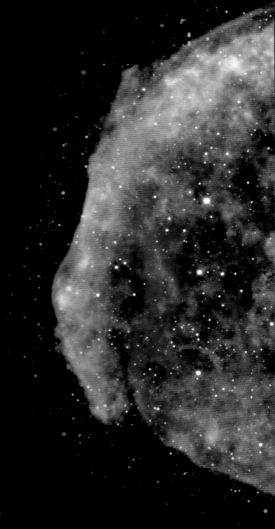

SUPERNOVA REMNANT
The Big Bang produced hydrogen, helium, and traces of lithium, but all the heavier elements of the Universe were formed within gigantic stars—the heaviest came from their supernova explosions.

At **500,000 years old**, the Universe contained the seeds of galaxy clusters among its filaments and voids.

At **1.3 billion years old**, individual galaxies had formed within crowded filaments that were rich in dust and young stars.

At **13.8 billion years old** (the present), the whole Universe has taken on a galaxy-supercluster structure.

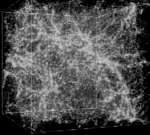
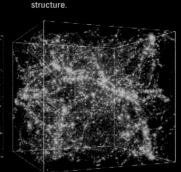

SIMULATING A GROWING UNIVERSE
Astronomers have used complex computer simulations to model the formation of galaxies and galaxy clusters in the Universe. This image shows the result of a simulation of the beginning of the Universe, from just after the Big Bang to 13.8 billion years after its occurrence (in other words, to the time we live in). As suggested by the simulation, astronomers have found that today's Universe contains vast filaments of galaxy clusters networking across space, as well as large, apparently empty regions. The simulations also consider the effects of the mysterious substance known as "dark energy" on the formation of galaxies.

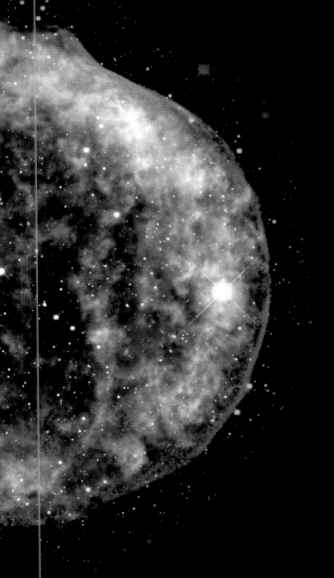

The first galaxies

Galaxies are huge collections of stars, and it is thought that the first galaxies began to form sometime around 400 million years after the Big Bang. These early groups would not have looked anything like "modern-day" galaxies such as the Milky Way. Observations made with orbiting telescopes such as the Hubble Space Telescope show that the earliest galaxies were small and irregularly shaped. Studies suggest that these early galaxies underwent vigorous star formation and that they were the building blocks of the galaxies we see today. Over time, these smaller galaxies collided and merged together to form much larger galaxies full of structure, like the ones near us in the Universe that we are able to observe today with amateur telescopes.

LOOKING BACK THROUGH TIME
The Hubble Ultra Deep Field (HUDF) image was obtained through a million-second-long exposure, and it shows some of the earliest and most distant galaxies ever observed. These little red smudges of light, lying beyond the much brighter nearby galaxies, were some of the first galaxies to emerge after the Big Bang.

THE NAKED-EYE SKY

Our eyes can make out many different objects in the night sky, including bodies in the Solar System, stars in our galaxy, and even neighboring galaxies.

What do our eyes reveal?

On a clear evening, the night sky appears as a sparkling dome of stars stretching from horizon to horizon. Away from the lights of towns and cities, a deeper level of this majestic nightscape is revealed, with the myriad stars of our galaxy, the Milky Way, creating a glowing, misty band across the sky. There are transient characters, too, in this nightly show: the fleeting streak of a shooting star or the occasional delicate shimmer of an aurora. Then there are the planets, our companions in space. They slowly wander across the background firmament as they orbit the Sun. The Moon is frequently present in the night sky. Its bright light can often wash out the fainter stars, but it is itself a fascinating object to observe, covered in smooth, dark "seas," as well as cratered and mountainous regions.

THE SPECTACLE BEGINS
There is no better way to enjoy the riches of the night sky than with the naked eye. There is so much that needs no equipment to be seen.

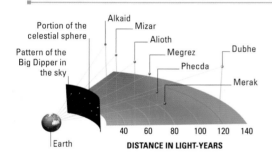

Portion of the celestial sphere

Pattern of the Big Dipper in the sky

Earth

Alkaid

Mizar

Alioth

Megrez

Phecda

Dubhe

Merak

DISTANCE IN LIGHT-YEARS
40 60 80 100 120 140

LINE-OF-SIGHT EFFECT
The patterns we see in the stars are determined by our line of sight. From Earth, we see the Big Dipper, but the stars would form a different pattern viewed from elsewhere in the galaxy.

THE ZODIAC
The zodiac is the ring of constellations, rich in myth and symbolism, through which the Sun appears to pass.

The constellations

Over time, humans have invented narratives to explain the patterns seen in the stars. These patterns are said to represent characters or objects from a mixture of ancient and relatively modern mythologies. Today the night sky is divided up by astronomers into 88 official areas, known as the constellations, each with its own demarcated boundaries. Stars located in the same constellation are not necessarily near each other—these groupings are just a convenient way to identify a particular area of sky.

Our galactic neighbors

Most of the objects that can be seen unaided in the night sky are stars that lie within our own galaxy—the Milky Way—but we are not alone in our corner of the Universe. The Milky Way and over 30 nearby galaxies make up a galaxy cluster known as the Local Group. Among this group are several galaxies that, from a dark-sky site, can be readily picked out with the naked eye, appearing as small, misty patches against the background stars. Two of the most prominent naked-eye objects in the southern hemisphere are two galaxies in the Local Group close to the Milky Way: the Small Magellanic Cloud and the Large Magellanic Cloud.

ANDROMEDA GALAXY
The Andromeda Galaxy is the largest galaxy in the Local Group. It lies around 2.5 million light-years away from Earth.

LIGHT-YEARS AWAY

When we switch on a lamp, we instantly see the light. That's because the speed of light (186,000 miles/300,000km per second) is so great compared to the distance it has to travel to our eyes. But the distances in space are vast, so it takes time for the light to reach our eyes. Astronomers express distance in terms of "light-years"; one light-year is the distance a beam of light would travel in one year. This means that when we see the light from a distant object like a star, we are in fact looking at light that left it a very long time ago. What we see is not the object as it is now. Incredibly, we are instead looking back in time.

Our **home galaxy**—the **Milky Way**—is just one of many **billion** galaxies in the observable Universe.

The galaxy we live in

Our galaxy, the Milky Way, is a huge collection of between 200 and 400 billion stars. All of the stars we see with our naked eyes on a clear night belong to the Milky Way. The galaxy measures some 180,000 light-years across and around 2,000 light-years deep, and it is a spiral galaxy (see p.26). Recent studies suggest the Milky Way has two main spiral arms. They are composed of bright young stars, and it is within one of these spiral arms that our Sun is located. The spiral arms are laced with clouds of dust that can easily be seen in the night sky as dark lanes weaving across the galaxy's brighter regions. At its center, the galaxy has a large hub of older stars known as the central bulge. It is this bulge we see when looking toward the constellations of Scorpius and Sagittarius and the surrounding area. The disc of the galaxy and its spiral arms stretch out on either side of this region.

THE MILKY WAY

Our galaxy can be seen as a hazy band arching across the night sky. If we could look down on the galaxy from outside it, we would see its spiral arms as shown in the illustration below.

THE PLANETS

Long before the time of telescopes, ancient astronomers noticed that there were several bright lights that moved, over the course of many days, against the background stars. The ancient Greek name for these moving objects was "planets," meaning "wanderers."

The Solar System

The Solar System is the region of space that contains the Sun and all the celestial objects that fall within its gravitational influence. It consists of eight planets, a handful of dwarf planets, and a myriad of other smaller bodies including comets and asteroids. All of these orbit the Sun, a yellow star at the Solar System's center that has been shining for 4.57 billion years. Many of the planets have their own moons too, each an intriguing world in itself. Beyond the main planets lie the colder outer reaches of the Solar System. Here sits the Edgeworth—Kuiper Belt, a disc of small, icy, orbiting bodies. This merges further out in space with a vast cloud of comets, known as the Oort Cloud, which encompasses the entire Solar System.

JUPITER ENTERING THE CONSTELLATION OF VIRGO
The Sun appears to travel through the sky along a path called the ecliptic. As the planets' orbits all lie close to the plane of the Sun's equator, they too all move close to this line. The constellations they pass through are collectively called the zodiac.

Planets, dwarf planets, and asteroids

The objects that orbit the Sun are divided into three classes: planets, dwarf planets, and small Solar System bodies. A planet orbits only the Sun, and its mass is so large that its own gravitational pull is strong enough to compact it into a spherical shape, while also sweeping the region of its orbit relatively free of other objects. A dwarf planet is one that also orbits the Sun and is large enough to be roughly spherical, but is not large enough to clear the region of its orbit. Small Solar System bodies, such as most asteroids and comets, are too small for gravity to round off their irregular shapes.

JUPITER

URANUS

ASTEROID IN THE ASTEROID BELT

The Sun accounts for **99.8 percent** of the mass of the Solar System.

ORBITING THE SUN
The Solar System contains an incredible array of worlds, from huge gaseous leviathans, with resplendent families of moons and rings, to smaller more modest rocky worlds such as our own Earth.

THE FORMATION OF THE PLANETS

The planets we see today, including the Earth, formed around 4.56 billion years ago. The Sun and all the planets in the Solar System formed from a great cloud of dust and gas called the "solar nebula." As the Sun emerged from the nebula, it was encircled by a huge disc of debris known as a proto-planetary disc. Inside this disc, material began to clump together, and before long planet embryos known as "planetesimals" were formed. Some closest to the Sun combined into the terrestrial, inner planets—Mercury, Venus, Earth, and Mars—and these were made predominantly of rock and metal. Further out in the cooler regions of the infant Solar System, icy planetesimals coalesced to form planet cores. These then captured gas from the disc and, in the process, the gas giant planets were formed: Jupiter, Saturn, Uranus, and Neptune. Astronomers have seen similar processes occurring around other stars, where new planetary systems may be forming.

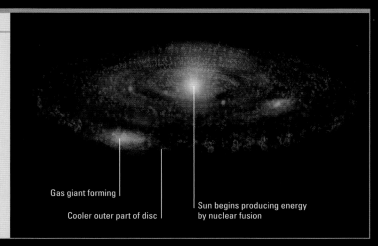

Gas giant forming

Cooler outer part of disc

Sun begins producing energy by nuclear fusion

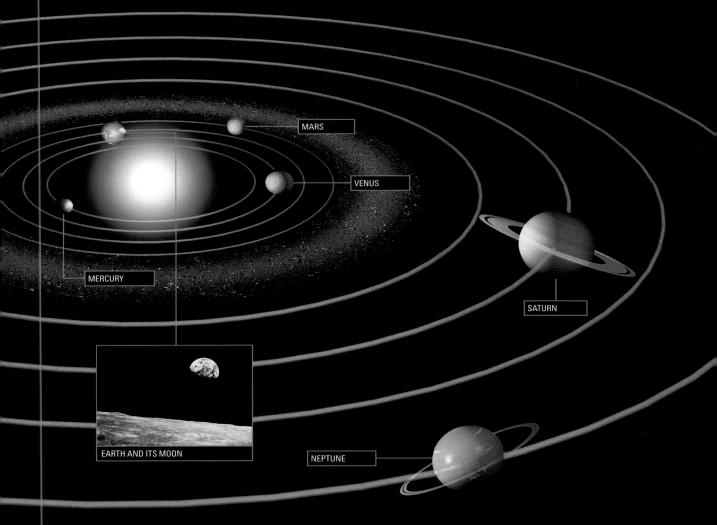

MARS

VENUS

MERCURY

SATURN

EARTH AND ITS MOON

NEPTUNE

WHAT IS A STAR?

The night sky is scattered with thousands of twinkling lights. Each one is a star—just one of hundreds of billions which inhabit our galaxy, the Milky Way. Studying the stars reveals how planets and even we humans came to be.

THE LIFE CYCLE OF A STAR
Stars pass through a series of stages during their lifetimes, the sequence and duration of which are determined largely by the star's mass.

A SKY OF MAIN-SEQUENCE STARS
In any typical view of the night sky, around 90 percent of the stars we can see are in the main sequence phase of their lives. Stars of a similar mass to the Sun spend around 10 billion years in the main sequence stage; those of smaller mass live far longer.

PROTOSTAR
Stars form from cold clouds of interstellar matter that collapse under gravitational disturbances.

COLLAPSING PROTOSTAR
As the temperature and pressure rise at the center of the collapsing cloud, material near the center flattens into a disc and nuclear reactions begin.

Smaller protostars do not achieve nuclear reactions, and these "failed stars" are known as brown dwarfs.

SUCCESSFUL STAR
As nuclear reactions get under way, young stars exhibit unstable behavior such as rapid rotation and strong winds. Their poles eject jets of material.

STABILIZING STAR
In stars with a mass of more than 0.08 solar masses, nuclear reactions start and the protostar becomes a star. The disc begins to cool and planets form around it. The star enters the main sequence phase of its life.

The night's light

A star is a huge ball of gas composed mostly of hydrogen. It is a sphere so enormous and massive that the temperatures and pressures at its center are large enough to fuse hydrogen nuclei together in a process known as nuclear fusion. This process releases vast amounts of energy, including the light we see when we look up and see a twinkling star. As a star fuses hydrogen in its core, it converts it into helium, and through further reactions it goes on to create progressively heavier elements—stars are responsible for creating all the elements of the Universe that are heavier than helium, such as oxygen, nitrogen, and iron.

TWINKLING LIGHTS
When the light from a star enters Earth's atmosphere, it encounters undulating air movements. These constantly bend and distort the light, making the star seem to "twinkle" to those observing from Earth.

A star is born

Stars are born inside great clouds of gas and dust called nebulae. The colder these clouds are, the more they are liable to react to a gravitational disturbance, such as shockwaves from a supernova. Denser knots of gas form and collapse, becoming denser and denser, until they form "protostars". Eventually the temperature and pressure at the core of a forming protostar become so high that nuclear fusion can take place. When this happens the star begins to shine, and it emerges from its maternal nebula to become a fully fledged star. It will continue in this state, known as the main sequence phase, for most of its life.

Our own **Sun** is a dwarf star, located 93 million miles (150 million km) away, which is now in the main sequence phase of its life.

Stellar death

Even though it is huge on the scale of the planets in our Solar System, our Sun is not very massive; there are stars over a hundred times its mass elsewhere in our galaxy. The mass of a star is crucial, because it determines how it behaves during its life and what happens to it when it dies. Massive stars convert their hydrogen fuel extremely rapidly, so their lives are short—perhaps only several million years—and they die quickly. Low-mass stars, on the other hand, use up their hydrogen much more slowly and can live for billions of years; our Sun will live for around 10 billion years, for example. When they die, stars with a mass similar to the Sun slowly expand to become red giants, before collapsing to become white dwarfs at the center of planetary nebulae. Stars of a lower mass simply shrink over time and fade away to become dim dwarf stars. Very massive stars die more violently, exploding as supernovae. When this happens, some supernovae create neutron stars, while the most massive become black holes.

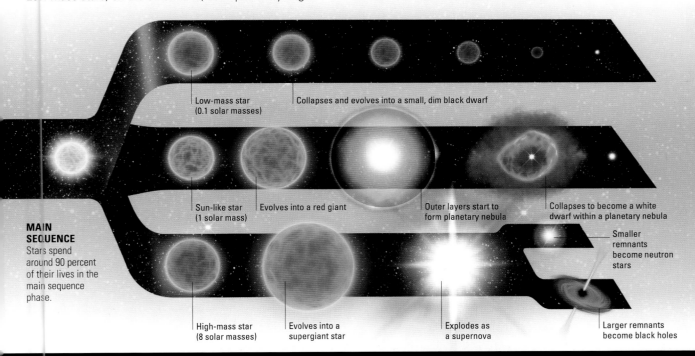

MAIN SEQUENCE
Stars spend around 90 percent of their lives in the main sequence phase.

Low-mass star (0.1 solar masses)
Collapses and evolves into a small, dim black dwarf

Sun-like star (1 solar mass)
Evolves into a red giant
Outer layers start to form planetary nebula
Collapses to become a white dwarf within a planetary nebula
Smaller remnants become neutron stars

High-mass star (8 solar masses)
Evolves into a supergiant star
Explodes as a supernova
Larger remnants become black holes

Evolution of a supergiant star

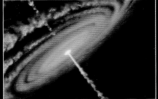

SUPERGIANTS
Supergiants are typically very massive stars that have run out of fuel to power their central nuclear furnaces and have swollen to enormous sizes. The largest known are around 1,000 times the diameter of our Sun. We can see many of these "supergiant" stars (glowing both blue and red) in the night sky. The bright star Antares (above) is a red supergiant in the constellation of Scorpius.

SUPERNOVAE
Eventually fusion reactions cease in the supergiant and there is not enough pressure from within the star to counter the force of gravity. The star collapses —the iron core collapses into a dense core of neutrons, and when the outer layers of the star impact on the rigid core they rebound back into space at tremendous speed, exploding and releasing vast amounts of energy.

NEUTRON STARS
If the stellar core that remains is less than 3 solar masses, it becomes a neutron star. These are extraordinarily dense stars made almost entirely of neutrons; they spin extremely quickly and possess very powerful magnetic fields. Some emit two beams of radio waves from their magnetic poles, which sweep across space like searchlights; these are known as "pulsars."

BLACK HOLES
If the stellar core was larger than 3 solar masses after the supernova explosion, the collapse cannot be stopped and an object known as a "black hole" is formed. This is a region in space that now has such a strong gravitational field that it draws in everything—even light. Because of this, astronomers can only observe the influence of black holes, not the holes themselves.

OBSERVING STARS

The colors, brightness, and size of stars allow astronomers to discover more about their evolution and the Universe they live in.

Brightness and color

When we look up at the stars, they do not all appear the same color, and they have varying degrees of brightness. A star's perceived color depends on its surface temperature, so some glow with a reddish hue (the cooler ones), while others sparkle a brilliant blue or white (the hottest ones). A star's perceived brightness depends on its actual brightness, as well as its distance from us. Some extremely bright stars look faint because they are very distant. Conversely, a faint star can be close to us in space and so appear relatively bright in the sky. Astronomers compare star brightness by establishing each star's "absolute magnitude." This is the magnitude (or brightness) a star would have if it were placed at a set distance from us (usually 32.6 light-years). "Apparent magnitude" is the magnitude the star appears to have in our night sky. So it is the differing apparent magnitudes of stars that are of real interest to amateur stargazers.

COLORS OF THE NIGHT SKY
Stars appear differently colored according to their temperature. Massive white-hot stars have short lives, so those we see are always young. Only cool stars live to a great age, so all old stars are red—though young stars can be red, too.

The Hertzsprung–Russell Diagram

The Hertzsprung–Russell (or H–R) diagram shows the relationships and differences between stars. Each star is plotted according to its surface temperature and luminosity (the amount of energy it radiates, commonly referred to as its magnitude or brightness). On the temperature scale, "0" sits to the right rather than to the left on the horizontal axis—so stars of higher temperatures are plotted to the left of the diagram. Stars to the upper left are hot and bright; stars to the lower right are cold and dim.

SIZE AND BRIGHTNESS
Stars that are inherently relatively faint (such as Sirius, the brightest star in the night sky) can appear brighter than more luminous ones because they are much closer to us.

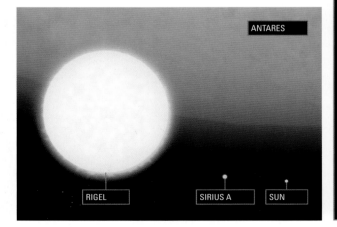

THE HERTZSPRUNG–RUSSELL DIAGRAM

When astronomers plotted stars on the H–R diagram, they found that they fall into groups according to the different stages of their lives, such as red giants or white dwarfs. Our Sun sits in a diagonal strip across the diagram that represents stars in the main sequence phase (see pp.16–17).

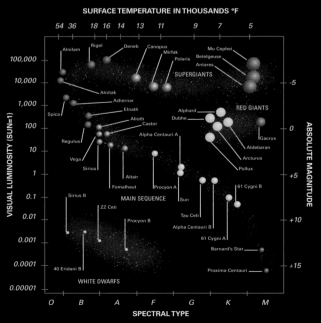

MULTIPLE STARS
Albireo (Beta Cygni) can be observed easily sitting among the Milky Way's stars at the head of the constellation of Cygnus. Even a small telescope will show the two main stars of this multiple star system shining blue and yellow.

Binary and multiple stars

It is thought that most stars in the Universe are accompanied by nearby companions. A pair of stars that are close to each other in space and that orbit a common center of mass (called a barycenter) is known as a binary star system. A star may even have more than one other companion in space. Many of the lovely multiple star systems in the night sky can be seen with even a small telescope, such as the gorgeous blue-and-yellow Albireo system (shown left) and Epsilon Lyrae (the "Double-Double" star system of four stars) in the constellation of Lyra. Optical double stars are stars that appear close to each other in the night sky (due to our line of sight) but that are not really bound to each other by gravity. A classic example of an optical double is the pair of stars Alcor and Mizar in the handle of the asterism of the Big Dipper. They appear to be near each other but are in fact widely separated in space.

Variable stars

Some stars in our night sky do not always shine at a constant brightness. This is not always obvious and can be difficult to observe, because the brightness (apparent magnitude) of these variable stars can fluctuate over the course of days, months, or even years. There are several possible reasons for this varying magnitude, from changes within the star itself to the dimming effect of a companion star passing in front of it.

Amateur and professional astronomers study variable stars for many reasons, but especially because they can provide clues as to how certain stars behave during key stages in their evolution. Variable star research is one of the areas of astronomy where amateurs, who make regular observations of variable stars, can really help the professionals. By recording the brightness of a variable star over an extended period, astronomers can then plot a graph known as a "light curve" (see diagrams, right). This is simply a curving line on a graph, mapping light levels from the dimmest point of a star to its brightest and then back to its dimmest. A light curve shows how the star's brightness fluctuates over time and helps determine the nature of the star's variability. The length of time between successive peaks or troughs on a light curve is known as a variable star's period.

CATACLYSMIC VARIABLES

Cataclysmic variables occur in a binary star system of a white dwarf and a cool star in its main sequence phase. Typically the white dwarf rips material off the other star onto itself, and this ignites in a large thermonuclear explosion, causing its brightness to rise spectacularly and creating a "nova"—a newly visible star. This image captures the explosion of Nova Cygni in 1992, which exploded so brightly that it was visible to the naked eye.

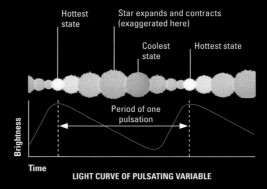

Hottest state
Star expands and contracts (exaggerated here)
Coolest state
Hottest state
Period of one pulsation
Brightness
Time
LIGHT CURVE OF PULSATING VARIABLE

PULSATING VARIABLES
These stars are variable because their atmospheres swell and contract, causing their temperature to rise and fall. As the star contracts, it becomes brighter; as it grows, it becomes dimmer. There are two main types of pulsating variable: Cepheids, which follow a regular, predictable pattern (shown); and Mira-type variables, which may change their pulsation periods from one cycle to another.

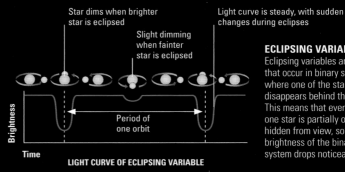

Star dims when brighter star is eclipsed
Slight dimming when fainter star is eclipsed
Light curve is steady, with sudden changes during eclipses
Period of one orbit
Brightness
Time
LIGHT CURVE OF ECLIPSING VARIABLE

ECLIPSING VARIABLES
Eclipsing variables are stars that occur in binary star systems where one of the stars regularly disappears behind the other. This means that every so often one star is partially or totally hidden from view, so the total brightness of the binary star system drops noticeably.

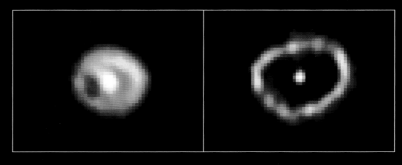

NEBULAE

When browsing the night skies with a telescope—
or even a pair of binoculars—it is easy to stumble
across a glowing nebula or cluster of sparkling
stars. These are some of the most beautiful
celestial objects that can be seen.

What are nebulae?

Most of the galaxy we live in is filled with an extremely tenuous, thin
gas consisting mainly of hydrogen and helium, and mixed with dust. It
permeates the space between the stars, so is known as the interstellar
medium. More dense clumps of gas and dust in space are known as
nebulae, and many of these are visible to us in the night sky.

Nebulae come in all different shapes and sizes, and from many different
sources. The huge nebulous regions can span many light-years across space;
the grand Lagoon Nebula in the constellation of Sagittarius, for example, is a
staggering 100 light-years in diameter. Some
nebulae are places where stars are born,
whereas others are the leftover
remnants of stellar death. Nebulae tell
a story of billions of years of stellar
life and death. For this reason they
are extremely interesting objects
for both amateur and professional
astronomers to observe.

TELESCOPIC VIEW
Through a telescope, a bright emission nebula will
appear as a faintly glowing region of sky, as shown
in this sketch of a telescope view of the Lagoon Nebula.

Stellar nurseries

Some nebulae in the night sky are regions within our galaxy's spiral arms
where stars are emerging. These so-called "star-forming regions" are
clouds of gas (mostly hydrogen), parts of which may glow a distinctive
ruby-red color. When disturbed they sometimes fragment into
smaller clouds, from which new stars are formed (see
pp.16–17). Many star-forming regions, such as the Orion
Nebula (M42) and the Lagoon Nebula (M8), are stunning
sights through a telescope. Often these stellar nurseries
have a cluster of bright, hot young stars embedded within
them (which they have given birth to), making them doubly
interesting to look at.

THE EAGLE NEBULA

The Eagle Nebula (inset) is a spectacular scene of star formation and has been
scrutinized by astronomers for centuries. It was the subject of the now-iconic
Hubble Space Telescope image that was subsequently dubbed the "Pillars of
Creation." This incredible image shows the towering clouds of gas and dust at
the nebula's heart that are giving birth to a new generation of stars.

Types of nebulae

EMISSION NEBULAE

Emission nebulae are clouds of gas that are being made to glow by a nearby source of radiation—usually a star or cluster of stars. Quite often emission nebulae are star-forming regions, such as the vast Lagoon Nebula (M8, pictured above), where clusters of emerging and young stars sit nestled inside the vast glowing expanse of the nebula. As the maternal cloud of gas is buffeted by harsh radiation from the hot young stars within it, it begins to glow. It is this emitted light that we see on Earth through our telescopes. Many emission nebulae can be seen easily on a clear dark night with little more than a good pair of binoculars or small telescope.

DARK NEBULAE

Against the backdrop of a bright nebula or a rich star field, a dark nebula appears as a silhouetted swirling mass. Dark nebulae are cold clouds of gas and dust that absorb light from stars or a bright nebula behind them. The iconic Horsehead Nebula in the constellation of Orion (pictured) is perhaps the most famous dark nebula of all. It has a total mass of about 300 times that of the Sun. Within the dark cloud from which it "rears" is a scattering of young stars that are in the process of forming. The Coalsack Nebula, in the constellation of Crux in the southern hemisphere, is another excellent example. It blots out the light from a dense Milky Way star field that lies behind it and is easy to see with a pair of binoculars.

REFLECTION NEBULAE

Most nebulae have tiny particles of dust inside them, which can reflect light from a star nearby. The density of the dust surrounding young stars is sufficient to produce a noticeable optical effect, and nebulae that appear to shine due to this reflected light are therefore called "reflection nebulae." Because the dust inside these nebulae is more effective at scattering the shorter-wavelength blue light, these nebulae usually appear pale blue in color. The stunning wispy streaks around the bright star Merope, in the Pleiades star cluster (pictured), show the characteristic blue color. While Merope is easy to see through a small telescope, the nebula around the cluster is much harder to observe.

PLANETARY NEBULAE

Planetary nebulae are heated halos of material shed by dying low-mass stars in the red-giant phase of their lives. They were named planetary nebulae by William Herschel in 1785, who thought they resembled planets. As a red giant dies, it expels the layers of gas in its outer atmosphere to expose the very hot core—a dense ball known as a white dwarf. Planetary nebulae are subject to contortion by magnetic fields and occasionally the influence of other stars. Their incredible shapes—such as the Dumbbell Nebula (shown above)— make them some of the most intriguing sights in the night sky. This stage of a star's life is relatively short, and eventually the nebula will disperse back into the interstellar medium.

STAR CLUSTERS

Nebulae are not the only glowing jewels to adorn our night skies. Star clusters provide stunning sights through all kinds of amateur equipment, from the simplest pair of binoculars to the largest telescope.

Open clusters

More than 1,000 sparkling open clusters of stars lie deep within the spiral arms of our galaxy. These clusters are formed when a group of stars is born inside a nebula. As they emerge from their maternal nebula, they sweep away much of the dust and gas around them, leaving the grouping of young stars that we see in our night sky as a glittering open cluster.

Open clusters are typically smaller than globular clusters (see below), with far fewer stars. Young open clusters are often dominated by a few massive, blue-white stars, but all open clusters contain stars with a range of masses. This is not indicative of age: the stars of an open cluster are all the same age, but have evolved to different stages according to their mass. Over time they will drift away from each other into the galaxy, so the cluster is always evolving.

OPEN STAR CLUSTER
The Butterfly Cluster is an open cluster full of relatively young, blue stars. It sits against the spectacular backdrop of the rich star fields of the Milky Way, in the constellation of Scorpius.

Globular clusters

A globular cluster is a densely packed ball of extremely old stars. The Milky Way possesses about 150 globular clusters, most of which are located in the halo (see top right). They range in size from leviathans such as Omega Centauri, which contains many millions of stars, to fairly small examples containing just 10,000 stars. Globular clusters are ancient objects, over 10 billion years old, that are largely devoid of gas and dust—there is no sign of star formation inside them, and no young stars. The cluster is tightly bound by gravity, and tends not to lose its stars, so the cluster ages with its constant constituents. The chemistry of the globular clusters in our own galaxy suggests that they represent remnants of the formation of the Milky Way.

GLOBULAR STAR CLUSTER
The globular cluster M15 in the constellation of Pegasus appears as a small glowing patch in an amateur telescope, but a larger one will show it to be a tightly packed sphere of stars.

An **amateur astronomer**, J. Abraham Ihle, was the first to discover a **globular** cluster—M22—in 1665.

Observing star clusters

The two types of star clusters are distributed differently around the Milky Way. Globular clusters are found in a halo around the main disc of the galaxy, and are often likened to bees swarming around a hive. Open clusters are found predominantly in or close to the spiral arms of the galaxy where there are rich clouds of gas and dust giving rise to star formation.

Through a small telescope or pair of binoculars, a globular cluster will appear as little more than a circular gray smudge against the background black of space. The larger the telescope's lens or mirror, the better the chance of distinguishing stars in the cluster. Open clusters are generally a little easier to observe than their globular counterparts. A good pair of binoculars is all that you need to see many open clusters in glittering detail. With a high-quality, small refracting telescope, open clusters can take on a stunning three-dimensional appearance.

CLUSTER DISTRIBUTION IN THE MILKY WAY

The difference in the locations of open and globular clusters within the Milky Way reflects their difference in age and orbit. Open clusters, formed from relatively young stars, are found within the galaxy's rotating disc. Globular clusters, made up of older stars, have independent orbits, mostly in the halo.

Central bulge   Halo

Spiral arm   Open clusters   Globular clusters

Notable clusters

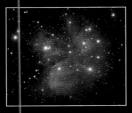

THE PLEIADES
Constellation: Taurus (see pp.126–127)
The Pleiades (or Seven Sisters) is one of the finest open clusters. It is easily visible with the naked eye from even a suburban location as a grouping of around 6 bright, blue stars. A simple pair of good binoculars will show that they are actually joined by a myriad of fainter stars.

M13
Constellation: Hercules (see pp.102–103)
M13 (also known as the Hercules or Great Globular) is by far the grandest of the northern hemisphere globular clusters, comprising several hundred thousand stars. A tiny circular smudge through binoculars, it becomes a glittering ball of stars through a telescope.

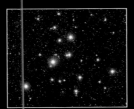

M39
Constellation: Cygnus (see pp.104–105)
M39 nestles against the dense background star fields of the Milky Way. This ancient open cluster is 200–300 million years old, and rests around 800 light-years from Earth. It is a very large but very loose cluster, and makes a fine sight through a small telescope from a dark sky site.

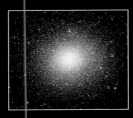

OMEGA CENTAURI
Constellation: Centaurus (see pp.142–143)
Omega Centauri is a huge globular cluster in the southern hemisphere constellation of Centaurus. Made of some 10 million stars, it is so large that some astronomers believe that it is not really a globular cluster but the core of a dwarf galaxy that has been devoured by the Milky Way.

THE HYADES
Constellation: Taurus (see pp.126–127)
Close to the Pleiades, the brilliant "V" shape of the Hyades is a very distinctive sight and makes up the head of Taurus, the bull. At only 150 light-years away, this stunning open cluster covers such a large area of sky that it is best scanned with a pair of binoculars.

47 TUCANAE
Constellation: Tucana (see pp.152–153)
This globular cluster, visible to the naked eye, is one of the finest in the whole sky. It sits in the southern hemisphere sky, near the Small Magellanic Cloud. With a magnitude of 4.9 it is a stunning sight through even a small telescope, appearing as a sphere of countless stars.

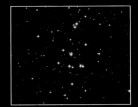

THE BEEHIVE CLUSTER
Constellation: Cancer (see pp.130–131)
The Beehive cluster, also known as Praesepe and M44, is a lovely, bright open cluster in the constellation of Cancer. With a magnitude of 3.7 the cluster is clear to the naked eye as a misty patch, and a pair of binoculars or small telescope reveals it to be made of many stars.

THE DOUBLE CLUSTER: NGC 869 & 884
Constellation: Perseus (see pp.122–123)
The Double Cluster is a pair of open clusters appearing close to each other in the sky. NGC 884 is 3.2 million years old, and NGC 869 is 5.6 million years old. Both are visible to the naked eye from a dark sky site and even a small telescope will show them in sparkling detail.

GALAXIES

A galaxy is a huge collection of stars, gas, and dust that is held together by gravity. Galaxies can be vast, with up to a trillion stars, or tiny, with just a few million.

What is a galaxy?

Galaxies take on many shapes, from huge round clouds of stars to intricate whirlpool shapes with defined arms full of stars. They may stretch from anywhere between a few thousand light-years wide, to more than 100,000 light-years across. Our Sun is just one star among the billions in the Milky Way, and this itself is just one among hundreds of billions of other galaxies. The Milky Way is part of a group of galaxies named the Local Group, which also contains the Andromeda (M31) Galaxy and about 45 other galaxies, including the Large and Small Magellanic Clouds. Clusters of galaxies are held together by gravity, and it is this same force that in turn links clusters of galaxies together to form superclusters. The Local Group is part of the Virgo Supercluster.

M74
The spiral galaxy M74 lies 32 million light-years away in the constellation of Pisces. Slightly smaller than the Milky Way, its arms are dotted with knots of glowing gas.

Types of galaxy

There are four main types of galaxies: spiral, elliptical, lenticular, and irregular, as shown below. The astronomer Edwin Hubble classified galaxies according to their different shapes. He prefixed ellipticals with the letter "E" and a number between 0 and 7 to indicate how far they deviate from a perfect sphere (0 being a perfect circle). Lenticulars are labelled S0. Spiral galaxies are broken into two groups: barred spirals are prefixed "SB" while spirals without bars have a simple "S". An extension of Hubble's classification divided the irregular galaxies into two types: "Irr I"—those with some hint of structure; and "Irr II"—those that appear to be completely disorganized.

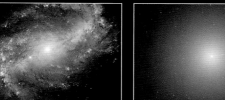

SPIRAL GALAXIES
Spiral galaxies account for around a third of the galaxies in the nearby Universe. Their central core of old stars is surrounded by spiral arms full of bright young stars. These galaxies are usually rich in regions of star formation, as they contain lots of gas and dust. Barred spirals, such as NGC 6217 (shown above), have a "bar" of stars across their central regions.

ELLIPTICAL GALAXIES
Elliptical galaxies appear as large "blobs" of stars and are usually full of old red and yellow stars. Unlike spiral galaxies, they contain hardly any star-forming nebulae. The largest of these galaxies are nearly perfect spheres. In the image above, a powerful jet is being ejected by a supermassive black hole at the center of the giant elliptical galaxy M87.

LENTICULAR GALAXIES
Lenticular galaxies typically possess a large, roughly spherical core of old stars surrounded by a disc of gas and stars, giving them an overall lens-like shape. The disc consists of stars and gas, as found in spiral galaxies, but these lenticulars—such as this one in the Coma Cluster—do not have spiral arms, young stars, or dusty, glowing star-forming nebulae.

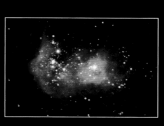

IRREGULAR GALAXIES
Irregular galaxies do not have a distinct shape. They typically contain lots of gas, dust, and hot, blue stars, but do not take any particular form. Some show signs of structure, such as central bars, and they often feature vast, pink, star-forming nebulae. NGC 2366 (shown) is a dwarf galaxy with a star-forming region 10 times bigger than that of the much-larger Milky Way.

Active galaxies

Observations of the distant Universe reveal some galaxies that emit vast amounts of radiation from their central regions. These are known as active galaxies. A supermassive black hole (see p.17) lies at the heart of each one, gorging on the rotating disc of dust and gas that is constantly falling into it. This releases tremendous amounts of high-energy radiation as well as creating two powerful jets which shoot out from either side of the black hole. Most, if not all, galaxies have black holes at their center, but in most cases these are inactive, so the material within the galaxies remains in a stable orbit around them.

There are four main types of active galaxy: blazar, Seyfert, quasar, and radio galaxies. They appear to have distinct features, but it is thought that their different appearances may only be due to the intensity of activity and the angle from which we are viewing them. The most distant objects yet discovered are the incredibly bright quasars.

QUASAR
The bright quasar designated as PG 0052+251 sits at the center of a spiral galaxy, around 1.4 billion light-years from Earth.

Observing galaxies

Galaxies are deep-sky objects, most of which lie millions of light-years from Earth, that present an interesting and rewarding challenge to amateur astronomers. Unlike stars, galaxies are spread out, and their surface brightness is relatively low. While a few are visible to the naked eye from a dark sky site, you will need at least a pair of binoculars and ideally a large telescope to see most of them. You'll quickly understand why they used to be mistaken for nebulae—their huge distance from Earth makes them appear as faint smudges of light in amateur telescopes, which is how they came to acquire the nickname "faint fuzzies." Some brighter galaxies may show signs of subtle structure, such as hints of spiral arms or dust lanes.

OBSERVING ANDROMEDA
The Andromeda Galaxy is the brightest galaxy in the northern hemisphere sky. At 2.5 million light-years from Earth, it is one of the most distant objects you can see with the naked eye.

DIVIDING LIGHT

Looking up into the night sky, it is sometimes hard to fathom what we are seeing. But by studying the light from space we can now determine the composition of a distant star, the age of a faraway galaxy, and what lies in a nearby nebula.

What is light?

Light is a form of energy known as electromagnetic radiation. It travels as waves, and can move through transparent materials, such as air, but also through empty space (unlike sound waves, which always require a medium). Visible light covers the range of electromagnetic wavelengths that are visible to the naked eye—they appear to us as the different pure colors of the rainbow, or color spectrum. The different wavelengths in a beam of white light can be separated by passing it through a prism, which splits the wavelengths to reveal their colors. Red light has a longer wavelength than blue light. Visible light, though, is only a small part of the total electromagnetic spectrum, which includes types of radiation our eyes cannot detect, such as X-rays at shorter wavelengths and radio waves at longer wavelengths. Radiation spanning the whole electromagnetic spectrum reaches us from across the Universe.

SHORTER WAVELENGTHS
Rigel is a blue-white star, emitting visible light at relatively short wavelengths; astronomers estimate its temperature to be over 21,600°F.

LONGER WAVELENGTHS
Betelgeuse's red color is due to it emitting light at relatively long wavelengths. Its surface temperature is only 6,150°F.

Packets of light

Electromagnetic radiation behaves like waves, but those waves come in tiny "packets" or particles of energy called photons. Their wavelength determines the amount of energy packed into the photon: the shorter the wavelength of the radiation, the more energy each photon possesses. X-ray photons are much more energetic than infrared photons, which is why they are potentially dangerous. Hotter objects, such as blue-white stars, not only emit more radiation overall than cool ones, such as interstellar dust clouds, but most of their emission is in the form of higher-energy photons at shorter wavelengths. The "intensity" of radiation reaching an astronomical detector, though, is set only by the rate at which photons arrive, not their energy levels. Red and white stars of the same brightness are equally intense, but the red star's photons carry less energy.

RESTRICTED VIEWS
Most amateur astronomers are restricted to observing visible light wavelengths.

OBSERVING OTHER WAVELENGTHS

Amateur astronomers mainly study the wavelengths of light visible to the naked eye. Yet for professional astronomers with complex observatories and telescopes at their disposal, observing the Universe at the many other wavelengths is a key tool for studying the cosmos. Earth receives seven types of electromagnetic radiation from space (listed right, from the lowest to the highest forms of energy).

SEE-THROUGH IMAGES
By observing at different wavelengths astronomers can look at different parts of the same object and even peer through thick clouds of dust and gas. The image on the left is an infrared view of the heart of the Milky Way, showing many stars in its dusty center. The image on the right shows the center of the galaxy in visible light, revealing the dust lanes that obscure the light from the stars beyond them.

Radio waves
Many of the most interesting objects in the Universe emit radio waves. These include pulsars, active galaxies, and vast clouds of hydrogen gas.

Microwaves
Perhaps the most significant type of microwave radiation originates from the Big Bang itself. The Cosmic Microwave Background (see p.10) is the ancient radiation left from the Universe's energetic birth.

Infrared
Stars, planets, and even our bodies emit infrared radiation. Observing infrared radiation from space is a good way to look into thick dusty nebulae, as it is able to pass easily through these objects to our telescopes, unlike visible light.

Visible light
Visible light is what we see when we look up on a clear night and see the light from the stars,

planets, galaxies, and other celestial objects streaming down toward us on Earth.

Ultraviolet
Hot young stars are very powerful emitters of ultraviolet radiation. Observing the Universe in ultraviolet light can reveal where new stars are being born.

X-rays
X-rays have more energy than ultraviolet rays, and can be found coming from very hot energetic objects, such as black holes and the superhot gas between galaxies in a galaxy cluster.

Gamma rays
These are the most energetic form of electromagnetic radiation, and carry the highest-energy radiation from space. Notable sources of gamma rays are gamma-ray bursts, which are thought to occur when a massive star explodes, or two neutron stars smash together violently.

A star's fingerprints
Light can tell us a great deal about the composition of the Universe we live in, because it is affected by the nature of its source, the distance it travels, its interaction with other matter, and so on. The different chemical elements and compounds in nebulae, stars, galaxies, and planetary atmospheres become clear to us by the unique "fingerprints" that they put on the light that comes from or through them. The light emitted by these objects can be broken down into a spectrum, within which we can find these chemical fingerprints.

Dark lines, called absorption lines, are found in a star's spectrum, and can be used to establish the constituents of its atmosphere. They are caused by photons being absorbed at certain wavelengths by atoms in the star's atmosphere. The composition of nebulae can be determined by bright lines within its spectrum, which are formed when the nebula emits light after being energized by a star.

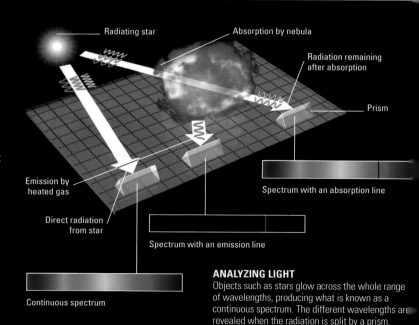

Radiating star

Absorption by nebula

Radiation remaining after absorption

Prism

Emission by heated gas

Direct radiation from star

Spectrum with an absorption line

Spectrum with an emission line

Continuous spectrum

ANALYZING LIGHT
Objects such as stars glow across the whole range of wavelengths, producing what is known as a continuous spectrum. The different wavelengths are revealed when the radiation is split by a prism. Absorption produces dark lines within the spectrum while emission leads to bright lines.

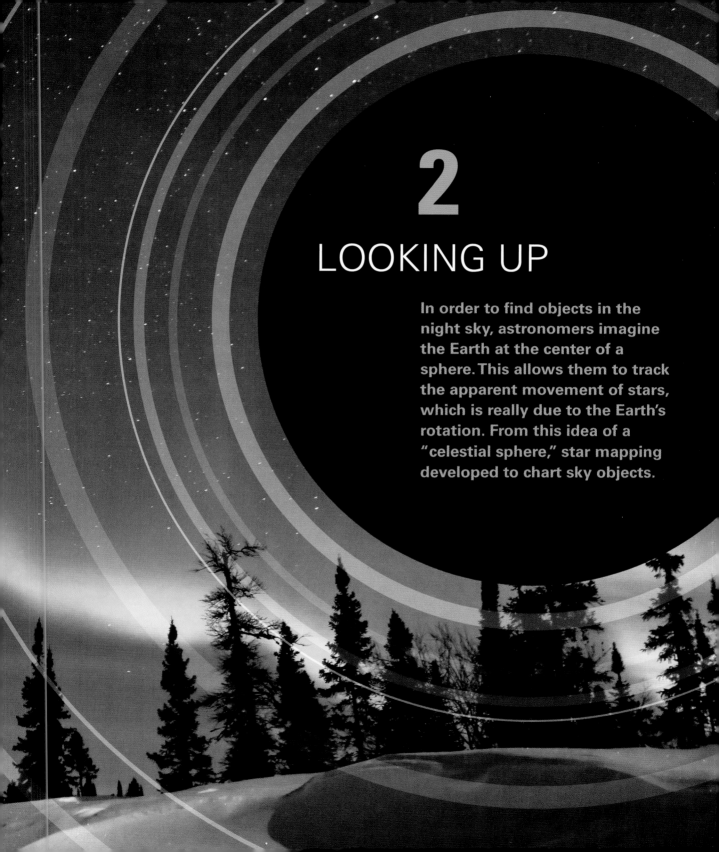

2
LOOKING UP

In order to find objects in the night sky, astronomers imagine the Earth at the center of a sphere. This allows them to track the apparent movement of stars, which is really due to the Earth's rotation. From this idea of a "celestial sphere," star mapping developed to chart sky objects.

THE MOTION OF THE SKIES

From Earth, it appears that the stars, planets, and Moon move around is. In fact, the movements that we observe result from the daily and annual movements of our own planet.

Our Earth-centered view

From the time of the ancient Greeks, Babylonians, and Egyptians, people have tried to make sense of the apparent cyclical movements of the stars, planets, Sun, and Moon. For the ancients, the Earth sat firmly at the center of the Universe, with the celestial bodies moving around it. These movements provided the bases for calendars that marked time for festivals and for farmers. It was not until the 16th century that the Polish astronomer Nicolaus Copernicus proposed that the motions of celestial objects could be explained without placing the Earth at the center of the Universe. Observations made by Galileo Galilei and others of Venus and Jupiter following the invention of the telescope in 1609, and Johannes Kepler's discovery of the laws of planetary motion, confirmed that the Earth is not fixed, but orbits the Sun along with the other planets.

MODELS OF MOTION
In the late 16th century, the Danish astronomer Tycho Brahe devised a variation of the Earth-centered Universe, in which the Sun, with the other planets going round it, orbits Earth.

Daily rotation

The most obvious movement of the Earth is its rotation on its axis, which can be visualized as a rod running through the Earth from the north to the south pole. Once every 24 hours, the Earth completes one full turn on its axis relative to the Sun, producing cycles of day and night: this rotation is what makes the Sun and stars appear to rise in the east, move across the sky, and set in the west. The time taken for the Earth to rotate once relative to the Sun (that is, from noon one day to noon the next) is called the mean solar day or the Earth's synodic day.

MOONRISE
The observed cycles of day and night arise because of the Earth's rotation around its axis.

THE EFFECTS OF LATITUDE

The Earth's rotation affects what can be seen in the night sky in different ways depending on the latitude of the observer. At the extreme location of the North Pole, the observer is effectively standing on the Earth's axis of rotation. He or she can see all the stars in the northern sky; the southern sky is obscured by the bulk of the Earth itself. Moreover, the stars appear to rotate around a single point that is directly overhead (near the Pole Star, Polaris) and never rise or sink below the horizon. If the observer moves to the equator, the view becomes very different; the Earth's rotation continually brings new stars into view as they rise, and hides others as they set. Between the poles and the equator, the view of the sky depends on the exact latitude.

The effect of the Earth's rotation can be clearly seen in photographs where the camera shutter is left open for periods of minutes or hours. The Earth's rotation makes stars appear to move through an angle of one degree of the night sky every four minutes of exposure. The resulting star trails in the photograph are a consequence of the Earth's rotation.

OBSERVER'S POSITION	APPARENT STAR MOVEMENT	OBSERVER'S VIEW	STAR TRAILS SEEN
NORTH POLE	The stars appear to circle around a point directly overhead. Their movement seems to be clockwise at the North Pole, and counter-clockwise at the South.		
MID-LATITUDE	At points between the poles and the equator, stars appear to rise in the east, cross the sky obliquely, and then set in the west. Depending on latitude, some stars are always in view, while others are always obscured by the Earth.		
EQUATOR	The stars appear to rise vertically in the east, move overhead, then set in the west. The Earth's rotation brings all parts of the sky into view for some time in each 24-hour period.		

Sidereal and solar days

While the Earth rotates on its axis once every 24 hours relative to the Sun, the Earth also moves around the Sun, taking one year to complete its orbit. In the time taken for Earth to spin once, it has also traveled a part of its way around the Sun. Astronomically, and somewhat confusingly, this leads to two different sorts of days: the usual 24-hour mean solar day (see left) and the sidereal day. Sidereal means "to do with the stars," and this "day" corresponds to one rotation of the Earth in relation to the stars. It lasts 23 hours 56 minutes and 4 seconds. Therefore, the time difference between the two "days" is nearly 4 minutes.

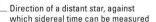

Direction of a distant star, against which sidereal time can be measured

PLAYING CATCH-UP
If the Earth did not orbit the Sun, but stayed rotating in a fixed place, the Sun and the stars would appear to us in the same positions every 23h 56m 4s. However, the Earth does orbit the Sun; after rotating once relative to the stars, it must turn for almost an extra 4 minutes before the Sun has circled our sky once.

Earth's orbit

Sun

Noon on first day

Earth's rotation

Second noon in solar time

Second noon in sidereal time (4 minutes earlier than solar time)

April 1 20.00   April 8 20.00   April 15 20.00

MOVING CONSTELLATIONS
The 4-minute difference between the sidereal and solar days means that the stars and constellations appear to shift slightly westward in the sky from one day to the next.

The Earth's orbit and the ecliptic

The definition of one year is the time it takes the Earth to orbit the Sun once. Of course, from our fixed Earth-centered perspective, the roles seem to be reversed and it appears as though the Sun travels around us. If you imagine that the Sun could leave an imprint in the sky at the same time of day, every day for a year, then after one year, we would see a complete circle around us. This circle is called the ecliptic, and it represents the Earth's path around the Sun. As well as the Earth and Moon, all the other major planets also orbit the Sun in the plane of the ecliptic—and appear to follow the same path around the Earth. This is because the Solar System formed in a large rotating disc of material and the larger objects have kept orbiting more or less on the same "level" or plane ever since. For ease, we use the Earth's orbital plane, the ecliptic, as the reference for the whole Solar System.

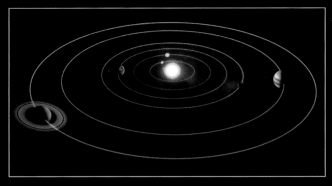

THE PLACE OF ECLIPSES
All the planets orbit the Sun on the same plane—the ecliptic. The word ecliptic means "place of eclipses"; lunar and solar eclipses can only occur when the Moon is at one of the two points in its orbit where it crosses the ecliptic.

TILTED PLANET
The Earth keeps the same angle of inclination throughout its orbit; the axis is parallel to itself at all points in its orbit.

Earth's axis ___ To Polaris

Ecliptic

23.5° inclination of Earth's axis

The Earth's tilt

Most globes of the Earth seen in schools or museums show our planet tilted over to one side; a line through the poles makes an angle of 23.5 degrees to the vertical. This is also the angle that the Earth's equator makes with respect to the ecliptic. In other words, as our planet travels around the Sun, it always "leans" over by this amount. If the axis is extended out from the North Pole into space it hits a point close to Polaris, the Pole Star. This star therefore seems to stay virtually in the same place all year (and beyond), with all stars appearing to rotate about it as the Earth turns. The same cannot be said of the Sun. From most locations, the Sun is higher in the sky in the summer and lower in the winter, and the amount of daytime changes with the seasons.

SHIFTING SUN
At mid-latitudes, the position of the Sun changes with the seasons. This composite image shows the Sun's path over the course of a day at three times of year. At the summer solstice the Sun spends the largest amount of time in the sky. At the winter solstice, daylight hours are at their shortest.

The zodiac

As the planets orbit the Sun along the ecliptic, they appear to an observer on Earth to move across a backdrop of constellations that lie near to the ecliptic. For ancient peoples, the movement of the planets was very significant, and was associated with myth, legend, and divination. The constellations through which the planets moved were collectively named the zodiac—from the Greek for "circle of animals." In today's skies, the planets still travel through the signs of the zodiac; some can occasionally be seen in other constellations near the ecliptic, such as Cetus, the Whale, and even Orion, the Hunter.

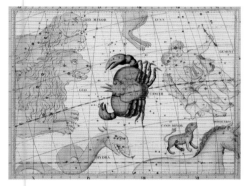

THE STAR SIGNS
In astrology, the zodiac—the ring of constellations along the ecliptic—is divided into 12 equal zones of longitude, each with its own sign.

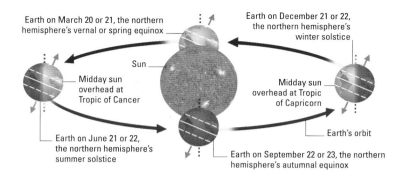

Earth on March 20 or 21, the northern hemisphere's vernal or spring equinox

Earth on December 21 or 22, the northern hemisphere's winter solstice

Sun

Midday sun overhead at Tropic of Cancer

Midday sun overhead at Tropic of Capricorn

Earth's orbit

Earth on June 21 or 22, the northern hemisphere's summer solstice

Earth on September 22 or 23, the northern hemisphere's autumnal equinox

SUMMER AND WINTER

Different hemispheres of the Earth tilt towards the Sun during the year; the one that is tilted towards the Sun is the hemisphere experiencing summer.

The seasons

The tilt of the Earth's axis in relation to the Sun accounts for the seasons that we experience. Around June 21 every year, the North Pole of the Earth is inclined as far as possible (23.5°) toward the Sun. This marks the start of summer in the northern hemisphere. At the same time the South Pole is inclined as far as it can be away from the Sun, marking the start of the southern hemisphere's winter. Astronomically this is the summer or winter solstice (depending on where you are on our planet). Six months later, the Earth has completed half of its orbit around the Sun and the North Pole is maximally inclined away from the Sun—this is the winter solstice in the northern hemisphere and the summer solstice in the southern hemisphere. Exactly half-way between the solstices are the equinoxes, when day and night are of equal length around our planet.

Axis of rotation
23.5° angle of tilt
Tropic of Cancer, 23.5°
Solar radiation
Tropic of Capricorn, 23.5°

WARMING LIGHT

The Sun warms the parts of the Earth that are inclined toward it more than those tilted away.

SUMMER SOLSTICE

SPRING/AUTUMN EQUINOX

WINTER SOLSTICE

THE SKY AS A SPHERE

A useful way to think about the position of the stars around the Earth is to imagine them embedded in a huge sphere, with the Earth at its center. The Earth spins from west to east within this sphere, completing one rotation in a day.

The celestial sphere

To understand how astronomers find objects in space, it is useful to think about how navigation works on Earth. Here, locations can be referenced according to a grid wrapped around the globe, which comprises the "horizontal" lines of latitude and "vertical" lines of longitude.

Latitude defines how far north or south a place is relative to the equator, while longitude defines how far a place is east or west of the prime meridian—a line that runs from pole to pole through Greenwich, London.

The same principle works for space. Imagine these lines of latitude and longitude projected out onto a "celestial sphere"—an imaginary sphere around the Earth on which all celestial objects appear to be located. On the celestial sphere, latitude is called declination (DEC), while longitude is known as right ascension (RA). All celestial objects can be located by knowing their DEC and RA coordinates.

OUR VIEW OF THE CELESTIAL SPHERE

All objects in space have their position on the celestial sphere, but an observer's view of those objects depends upon their location on Earth. A stargazer standing at the North Pole can see everything from the celestial equator, at 0° DEC, up to the point overhead, which is +90° DEC, close to the Pole Star; therefore, only the northern half of the celestial sphere can ever be seen. For an observer at the South Pole, the situation is reversed. A stargazer on the equator, however, can see all the way from -90 to +90° DEC, allowing a full view of the entire heavens over the year. An observer at mid-latitudes can see only part of the opposite hemisphere's celestial sphere depending on their exact latitude.

LIMITED VIEW
Your location will usually restrict what objects can be seen.

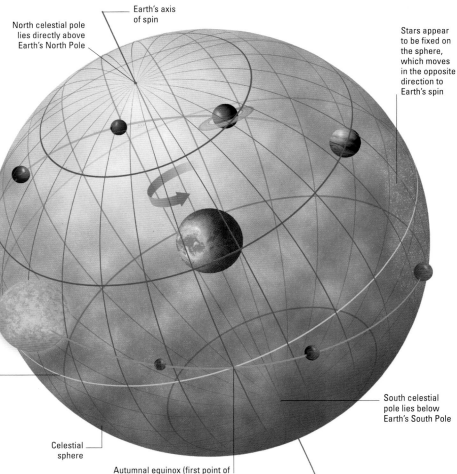

Earth's axis of spin

North celestial pole lies directly above Earth's North Pole

Stars appear to be fixed on the sphere, which moves in the opposite direction to Earth's spin

The Sun and planets are not fixed on the celestial sphere, but move around on or close to the circular path known as the ecliptic

Celestial equator—a circle on the celestial sphere concentric with Earth's equator

South celestial pole lies below Earth's South Pole

Celestial sphere

Autumnal equinox (first point of Libra), one of two points of intersection between the celestial equator and ecliptic

DEFINING POSITIONS

The notion of stars embedded in a sphere around the Earth is a relic of ancient Greek astronomy but remains a useful concept. It is convenient to think of stars as having celestial coordinates in the same way that objects on Earth have a unique latitude and longitude.

Understanding celestial coordinates

Of the two celestial coordinates, declination (DEC) is the easier to understand. Imagine projecting the Earth's equator onto the celestial sphere; this line is known as the celestial equator, and it is where declination is zero. Just as with latitude on Earth, declination uses degrees and minutes to measure the angle relative to the celestial equator. Declinations between the equator and the north celestial pole have positive values; those between the equator and the south celestial pole, negative.

Right ascension (RA) seems more complex because it uses hours and minutes rather than degrees for its measurement scale. One hour of RA across the celestial sphere is equivalent to 15° of longitude here on the Earth. The reason time is used instead of an angle is because in one hour the sky appears to turn through 15°. The line of zero RA, or the celestial meridian, is marked at the moment when the Sun, moving northward along the ecliptic, crosses the celestial equator. However, this position is not permanently fixed due to a long-term "wobble" of the Earth's axis, which is known as precession (see panel, right). This means that the entire celestial grid shifts very slightly each year: when coordinates are given for astronomical objects, they are technically only correct for a particular date.

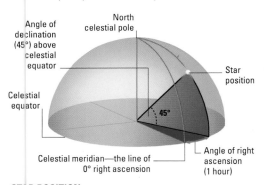

STAR POSITION
The above diagram shows how the celestial coordinates of a star are measured. The star above has a declination of +45° and a right ascension of 1 hour.

SEEKING STARS
A telescope fitted with an equatorial mount (see p.55) allows astronomers to locate stars if their equatorial coordinates are known. Two dials are used to set the telescope's DEC and RA to match the published

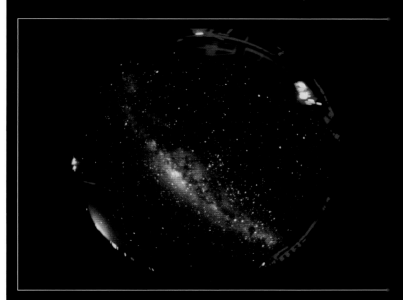

Altitude-azimuth coordinates

Using RA and DEC allows astronomers to precisely specify the location of an object in the sky. However, for more casual observation, it is more common to use a different system of coordinates called the altitude-azimuth (alt-az) system. Here, the horizon provides the frame of reference, and an object's location in the sky is described relative to it. Altitude is measured up from the horizon in degrees, while azimuth is the location around the horizon from true north, again measured using degrees. As an example, a planet may be located at 270° azimuth and 45° altitude. However, the azimuth location is often colloquially shortened by observers to a simple compass bearing, and altitude is often referred to simply as "up." So you may very well hear astronomers describing the position of a star as "in the west and 45° up."

VIEW HORIZONS
Alt-az coordinates use the horizon around the observer as their reference point, providing quick location of objects in the sky.

PRECESSION

Just like a spinning top, the Earth "wobbles" as it spins, changing the direction of its axis. In practice, this is of little consequence for the amateur astronomer because the wobble is extremely slow, taking 25,800 years to complete one cycle. However, it does mean that the Earth's relationship to the celestial sphere does change over time.

MOVING POLES
The Earth's axis points to the star Polaris today, but precession will make it point to Alderamin in 8000 or

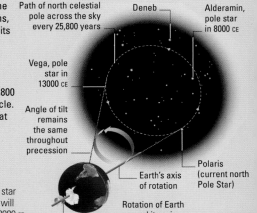

STAR TRAILS
As the Earth rotates around its axis, the whole sky seems to wheel around over our heads. Long-exposure photography records this motion as stars become arc-shaped trails centered on the celestial pole.

NAVIGATING THE NIGHT SKY

Knowing your time and location, and equipped with some knowledge of the relative sizes and brightnesses of celestial objects, you will be all set to start navigating the night sky.

Getting your bearings

As the Earth spins on its axis and moves around the Sun, the view of the stars from the Earth's surface changes quite quickly. Therefore the starting point for any astronomical observation is a knowledge of the date and time, as well as the latitude and compass orientation of the observing site. Once you have this information, it is a simple matter to use a planisphere, star chart, or computer software (see below and opposite) to show what should be visible and where, at the time of observing.

POINTING THE WAY
A simple compass is a useful astronomical tool. It tells you which horizon you are facing and can be used to roughly align a telescope mount with the night sky.

ANCIENT PATTERNS
People have named and used star patterns for centuries. The oldest constellations date back over 3,000 years, while the most recent were invented in the 18th century.

Using the constellations

The clearest signposts in the night sky are the constellations. These stellar patterns have long been used by sailors navigating the globe and by astronomers finding their way around the celestial sphere. Knowing how to recognize the major constellations and use them as signposts is just as important to the modern stargazer. Studying star charts can help with this, but there is no substitute for outdoor observation for getting a real sense of what the night sky looks like; stars often do not appear as bright (or faint) as expected, and constellations may appear much larger or more inconspicuous than anticipated. Start by getting familiar with some of the brightest constellations and use these as starting points to locate other, fainter, constellations and their stars. Before long, even the less prominent ones will become familiar, so that finding your way around the sky becomes a simple task.

Using a planisphere

A planisphere is a simple device that helps you quickly work out what lies where in the night sky on any given evening. It is an ideal companion for naked-eye astronomy and is especially useful when you are out in the field without access to a computer or a bulky star atlas. A typical planisphere consists of a circular piece of plastic with an oval-shaped "window", attached to a circular cardboard or plastic base printed with a star chart. The transparent oval window represents the whole hemisphere of the sky above the observer. The plastic upper disc can be rotated over the star chart beneath it, which changes the stars visible through the oval window. Planispheres are usable only within a certain range of latitudes, so be sure to buy the one appropriate to your location or you will get incorrect sky views.

PLANISPHERE
A planisphere is marked with points of the compass, date, and time. Stars of higher magnitude are represented by larger dots than fainter stars. The oval-shaped window shows the stars visible at a given time on a given night.

SETTING DATE AND TIME
To use the planisphere, turn the wheel so that the time of observation aligns with the date. Here the time and date are 11pm on 12 August.

ALIGNING AND VIEWING
Hold the planisphere over your head and turn it until it is aligned with the points of the compass—so that the North marker points to the northern horizon. The oval window now shows a map of the constellations visible above the horizon. The edge of the oval window corresponds to the horizon around you.

Using star atlases and software

A planisphere, or a set of basic star charts such as that on pp.210–233, is a good guide to the night sky but, by necessity, it compresses the whole of the sky into a small window or page. An alternative is a star atlas—a book containing more detailed charts of the whole of the celestial sphere. Some atlases plot the sky down to magnitude 6.5—ideal for binocular and naked-eye observing—while others plot stars several magnitudes fainter, making them more suited to telescopic observers. Some atlases add a great deal of other practical reference material; for example, dates of the principal meteor showers, lists of notable variable stars, and selected objects of interest to look out for on each chart.

In recent years, the principle of the planisphere has been extended into dedicated astronomical software—the computer-based planetarium program. There are numerous packages available, from relatively simple software that plots the positions of the stars, planets, and Moon for the observer's location, to extremely detailed programs with vast catalogs of stars and deep-sky objects. Some render the sky in photorealistic detail and can even provide the capability to control a telescope via the computer. With some programs you can also print out star charts, or switch to red night vision mode for use on a laptop in the field. Stellarium, a superb program that can be downloaded free from the Internet, is an ideal introduction to planetarium software.

PORTABLE PLANETARIUM
Sophisticated planetarium applications can be loaded onto smartphones. These display the view of the night sky after the user inputs details of location, time, and date.

Brightness and magnitude

Knowing how bright an object appears in the night sky is an obvious aid to identification and navigation. It is important to distinguish between an object's absolute magnitude—how bright it really is—and its apparent magnitude—how bright it appears to an observer on Earth. Apparent magnitude is the measure commonly used by stargazers.

Apparent magnitude differs from absolute magnitude because the further away a star is, the fainter it appears—just like a torch in your hand looks visually brighter than an identical one some distance away. So, Alpha Centauri has a higher apparent magnitude than Betelgeuse, because it is much closer to the Earth, but Betelgeuse has a higher absolute magnitude because it is much more luminous.

The scale used by astronomers to measure apparent magnitude is based on assigning a magnitude of zero to the star Vega: brighter objects have a negative magnitude value, and a rise in magnitude of 1 roughly corresponds to an increase in brightness of 2.5 times.

SELECTED APPARENT MAGNITUDES

OBJECT	APPARENT MAGNITUDE
The Sun	-26.7
The Moon	-12.6
Venus	-4.7
Mars	-2.9
Jupiter	-2.9
Mercury	-1.9
Sirius (brightest night-time star)	-1.4
Saturn	-0.3
Ganymede (moon of Jupiter)	4.6
Asteroid Vesta	5.3
Uranus	5.5
Faintest naked-eye objects	around 6.0
Neptune	7.7
Pluto	13.8

BRIGHT AND FAINT STARS
The brighter an object appears, the lower the value of its magnitude. Sirius, the brightest star in the night sky, shown here, has an apparent magnitude of -1.4. The Hubble Space Telescope has detected stars with a magnitude of +30.

Exploring by catalog

There are thousands of interesting objects in the night sky other than stars and constellations. Many of these objects have been observed, listed, and named, and today there are a number of catalogs that list these interesting nebulae, galaxies, star clusters, and more. Many astronomers use these lists as a "menu" for a night's observation.

By far the most famous of the object lists is the Messier catalog, containing over 100 interesting deep-sky objects. Objects in the catalog are denoted by the letter "M" for Messier—for example, M42 (the Orion Nebula) and M45 (the Pleiades). Such is the popularity of the objects in this list that some amateur astronomers engage in a Messier marathon. This involves attempting to observe as much of the Messier catalog as possible in one night. In addition to the Messier Catalog, there is the much larger New General Catalog (NGC), containing over 7,800 objects, and the Index Catalog (IC) of over 5,000 objects, both of which were created by Johan Ludvig Emil Dreyer in the 19th century. Some objects have both Messier catalog numbers and NGC numbers.

CATALOG NUMBERS
The reddish emission nebula of Orion has the Messier number 42. Below it, the blue nebula NGC 1977 is a reflection nebula, which gets its color from nearby young stars.

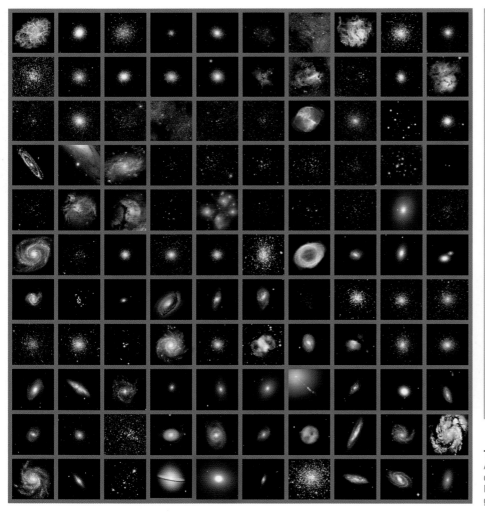

MESSIER AND MÉCHAIN

The majority of the Messier Catalog was compiled by the Frenchman Charles Messier (1730–1817) during the 18th century. Messier was an accomplished astronomer and expert comet hunter, discovering 21 comets during his lifetime. He began compiling his now-famous catalog in order to create a list of deep-sky objects not to be confused with comets. Messier and his colleague Pierre Méchain spent several years scanning the night skies discovering most of the objects that appear in the catalog. Today the Messier Catalog contains 110 objects representing a rich selection of the night sky's finest deep-sky sights.

CHARLES MESSIER

THE MESSIER CATALOG
All 110 Messier objects are shown in this composite image. They range from the Crab Nebula (M1) at top left to the dwarf elliptical galaxy M110 at bottom right.

Measuring sizes

Knowing the apparent size of a constellation or the apparent distances between objects is a great help in navigating the sky. These distances are measured in units called degrees—which express the apparent angle between the two objects. For naked-eye observation, our hands, held out at arm's length against the background of the sky, make very useful measuring devices.

HANDSPAN
A fully outstretched hand held at arm's length spans about 22° of the sky.

FINGER JOINTS
A side-on fingertip is about 3° wide; the second joint is 4°, the third joint 6°.

FINGER WIDTH
One finger at arm's length will cover the Moon, which is less than 1° across.

- 1 degree
- 90 degrees
- 360 degrees

Using hands is a great way to measure large angular areas of the sky, but things change when you look through a telescope, where the field of view is usually less than one degree. When it comes to smaller sizes, degrees are divided into 60 equal segments called arcminutes. These units may also be written in various other forms, such as minutes of arc, arcmin, or just the symbol ' (for example, 43'). For even smaller divisions, the arcminute is divided into 60 arcseconds, seconds of arc, arcsec, or the symbol ". When observing star clusters, nebulae, or the distance between double stars, arcminutes and arcseconds are the more common units.

ANGULAR DISTANCES

OBJECT OR DISTANCE	APPROXIMATE ANGULAR SIZE
Distance from the Pointers in the Big Dipper to Polaris	28°
Length of the Big Dipper	24°
Distance between the Crux Pointers	6°
Distance between the Pointers in the Big Dipper	5°
Your little finger at arm's length	1°
The Sun (average size)	32'
The Moon (average size)	31'
Distance between Jupiter and Ganymede (the brightest of its main moons)	6'
Resolution of the naked eye (this means the ability of your eye to split two objects that are as close together as this)	3' 25"
Maximum size of Venus	1'
Biggest crater on the Moon	1'
Smallest object your eye can see as a disc and not a point of light	1'
Maximum size of Jupiter	49"
Distance apart of the double star Beta(β) Cygni in the constellation Cygnus	30"
Smallest object of definite size that can be seen with a telescope on Earth	1"
Object size limit for the Hubble Space Telescope	0.1"

URSA MINOR

Pherkad

Kochab

Polaris (The North Star)

1

START AT URSA MAJOR (THE BIG DIPPER)

Alkaid

Mizar

Alioth

Megrez

Phecda

Dubhe

Merak

Starhopping

A deservedly popular way of finding your way around the night sky is "starhopping." As its name suggests, this involves using known stars or a group of stars to point the way to other stars and celestial objects. This method works just as well whether you are observing with the naked eye or through a telescope or binoculars. Starhopping is not only a useful practical method for navigating from one object to another, but also a good way to become familiar with different bright stars and significant constellations over time.

You can embark on a number of starhops from constellation to constellation (see pages 84, 106, 114, 136, and 140) or on a much smaller scale—useful when trying to locate faint objects, such as galaxies, in the eyepiece of a telescope. To do this, first locate a bright star that is near your target object using a low-power eyepiece. Then use a detailed star chart or printout from a planetarium program to move from star to star until the target object is in the field of view.

STELLAR SIGNPOSTS
The starhops described in this book, such as that shown here from the Big Dipper to Polaris (see p.84), are an excellent introduction to this interactive way of getting to know the night sky.

OBSERVING CONDITIONS

What is visible in the sky on any given night depends on a variety of factors, from your viewing location to light pollution, air turbulence, and particles in the atmosphere.

CLEAR SPACE
Orbiting the Earth at an altitude of 347 miles (559km), the Hubble Space Telescope allows observation well above the turbulence and pollution of our atmosphere.

Real-life views

Few people fail to be awed by the images of distant galaxies and nebulae returned by space telescopes, such as the Hubble Space Telescope, and professional observatories on Earth. Exquisite views are still available to an amateur with a modest telescope, but it is sensible to be realistic about what you will and will not see. The views that can be achieved will depend on the type of telescope used and the viewing conditions, which are explored on these pages.

HIGH EXPECTATIONS
The best viewing conditions on Earth can be had at high-altitude locations, where observers are elevated above layers of atmospheric pollution. Sites such as Mount Pinos, California, shown above, once attracted hundreds of amateur astronomers, but have since been compromised by development that has added to light pollution (see opposite).

WHAT CAN YOU SEE?

As a rough guide, an amateur observer equipped with a good 300mm reflecting telescope (see p.54), under fine observing conditions, could expect to see the celestial objects listed below:

THE MOON	Craters, seas, mountains, and other details on the Moon can be seen in fine detail.
THE PLANETS	Jupiter appears around the same size as some of the medium-diameter craters on the Moon; its cloud bands, Galilean Moons, and perhaps the Great Red Spot are visible. Mars appears the same size as a baseball seen from a distance of 10ft (3m). Details on its surface are visible when Mars approaches the Earth (see p.180). Uranus looks like a tiny greenish dot.
STARS	A telescope makes stars brighter, but they still appear just as points of light.
DEEP-SKY OBJECTS	Galaxies and nebulae appear as smudges of light; star clusters are bright and jewel-like.
COLOR	Only some planets and a few larger stars, such as Betelgeuse and Rigel in Orion, show any color. The human eye is not sensitive enough to reveal color in nebulae: the stunning images of these objects seen in this book and elsewhere are made using CCD cameras and very long exposures (see p.66).

Reducing the effects of the atmosphere

An often-overlooked factor that influences how an object appears to an observer is its altitude in the sky; indeed, stars may appear around three magnitudes fainter near the horizon than when directly overhead. The reason for this is that when a star is overhead at its highest, its light needs to pass through a thinner section of the atmosphere than when it is near the horizon. The atmosphere contains dust and particles, especially near towns and cities, so it absorbs some of the light from the star; and turbulence in the atmosphere also distorts the view of the star. For the best views of objects, it pays to wait until they reach their highest point in the sky.

OVERHEAD ADVANTAGE
Observing an object when it is directly overhead reduces the thickness of atmosphere through which its light must travel, and so improves the view.

Dark skies

Even the casual observer knows that the clearest views of objects in the night sky can be had in the countryside, well away from the glow of light from large towns and cities. This unwanted glow—called light pollution—is a problem for astronomers because light cast upward into the atmosphere from streetlights, billboards, and other sources reduces the contrast between the dark background of space and objects. Put another way, the darker the sky, the fainter the objects that can be picked out. The faintest objects that can be seen with the naked eye in city skies have a magnitude of around 3.5, while in dark-sky sites, experienced astronomers can pick out objects of magnitude 6.5.

Brighter objects, such as the Moon, the planets Venus, Jupiter, and Saturn, and the most prominent constellations, are good targets for observation even in light-polluted urban skies, but astronomers seeking deep-sky objects should seek out dark skies a good distance away from settlements.

Another source of light pollution relevant to deep-sky observers is the Moon, especially when at its large gibbous or full phases. A full Moon reduces viewing contrast and also destroys the dark adaptation of the human eye (see p.48). When moonlight enters our atmosphere, it gives the sky a bright, pale-blue appearance that washes out all the faint objects in a large area all around the Moon, so the best time to observe is before the Moon has risen, after it has set, or during a new Moon.

GOOD LIGHTING
Some city governments are becoming more aware of the problem of light pollution and installing better street lighting, which minimizes the amount of light spilling upward into the night sky and reducing visibility.

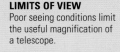

LIMITS OF VIEW
Poor seeing conditions limit the useful magnification of a telescope.

SEEING CONDITIONS

"Seeing" is a word used by astronomers to describe the steadiness of the atmosphere. When using a telescope to observe at high magnifications, the ability to make out fine details is limited by turbulence occurring in the telescope (local seeing) and in the atmosphere. When the seeing is bad, astronomers are limited to usable magnifications of around 100x, but in good seeing conditions magnifications of more than 300x can be used to pick out detail in objects.

Seeing can be quantified using the Antoniadi Scale, which was developed for use for the Moon and the planets. It is a simple five-point scale using the Roman numerals I through to V. On the scale, I is the best, being perfect seeing without a quiver; II represents slight disturbances, but calm air for several seconds at a time; III is acceptable seeing; IV means the viewing has a constant tremor and is considered poor; while V is extremely bad with severe undulations that do not allow features to be made out properly.

Turbulence and transparency

Local air turbulence can have a profound effect on viewing. If the Sun has been beating down on a concrete yard all day, the yard will stay warm long into the night, heating and stirring up the air overhead. A telescope observer in the yard will effectively be looking at the night sky from the bottom of a pool of swirling air, which will cause a distorted view. The remedy is to pick a spot that has been shaded during the day to minimize temperature differences. Similarly, it is best to avoid setting up a telescope indoors or to observe an object that is over someone's house: heat rising from the house mixes with cooler air above, causing turbulence that affects viewing.

Turbulence also works on a smaller scale. When a telescope that has been kept indoors in the warmth is taken out into a cold night, variations in temperature within the instrument can cause the images in the eyepiece to be turbulent and shimmery. For the best views, put the telescope outside for an hour or so before observing to make sure it is at the same temperature as the surrounding air.

ATMOSPHERIC TRANSPARENCY
A cloudy sky will make observing impossible, but a hazy sky, heavy with water vapor, will also make viewing difficult. A bright halo around the Moon indicates that atmospheric transparency is poor.

3
OBSERVING AND RECORDING

Using the right binoculars or telescope will make all the difference to the success of your stargazing. Once you're used to working with them, you might want to explore ways of recording your observations, such as astrophotography or sketching what you see.

PREPARING TO OBSERVE

Exploring the night sky does not require any specialist equipment—just curiosity, patience, and a little sensible preparation.

Do I need a telescope?

Astronomy can be enjoyed by anyone with reasonable eyesight. Without any optical aids, you can trace the shapes of constellations; follow the paths of five planets across the sky; pick out star clusters like the Pleiades; spot meteors; and observe the Orion Nebula (M42). Of course, a decent pair of binoculars or a good telescope will allow a greater choice and let you see fainter objects, but many astronomers would argue that you should start with naked-eye observation to get familiar with the night skies before investing in optical equipment.

NAKED-EYE VIEW
For some, there is no better way to experience the riches of the night sky than with the naked eye: many and varied objects do not need any equipment to be seen.

Getting ready

The best nights for astronomy are the clearest—and often the coldest! Being prepared for lower temperatures will make a night's observing far more comfortable. A warm hat, fingerless gloves—that let you retain your dexterity—and a warm, windproof jacket are essential accessories, and a container of hot tea or coffee is always appreciated. Check that you have all you need for your observing session—if you have to go back indoors, your eyes will lose their dark adaptation (see below). Astronomy involves standing outside for long periods, or even sitting or kneeling on damp ground to see through a sometimes awkwardly angled eyepiece. Comfortable shoes and waterproof pants should be high on your list of must-have observing accessories.

NIGHT DRESSING
Layers of warm, comfortable clothes are the best choice for a night's observing. Dressing warmly will mean that you can observe for much longer.

Dark adaptation

Under dark skies, your eye can distinguish stars of around magnitude 6; however, if you allow your eyes around 30–60 minutes to adapt to dark conditions, you should be able to see stars of around magnitude 6.5 under perfect conditions. This phenomenon, called "dark adaptation," occurs because our eyes contain two types of light receptors: cones, which work best in daylight; and rods, which work best in low light conditions. It takes the rods some time to adjust to new light intensities. Another trick to help view faint objects is averted vision. This involves looking slightly to one side of an object (whether with the naked eye or through a telescope); the light from the object is made to fall on the more sensitive rod cells in the eye.

SAFE LIGHT
A red LED light or a flashlight covered with a red filter preserves night vision, and is a useful accessory.

GETTING INVOLVED

Astronomy need not be a solitary occupation. One of the best ways to pick up knowledge is to join an astronomical club; most hold regular meetings with practical demonstrations of equipment, workshops, and talks by visiting scientists. Some clubs have access to their own observing sites, or even operate their own observatories that house large telescopes. Astronomical clubs may also organize "star parties," where astronomers gather to observe for several nights under dark skies. They are a great way to meet other astronomers and a fantastic opportunity to look through a huge variety of telescopes.

STAR PARTY
All around the world, astronomers gather together at dark sky sites, well away from light pollution, to share their passion for the night sky.

Making observations

At a time when inexpensive digital cameras—teamed with telescopes— can produce vivid images of the sky (see p.64), it may seem anachronistic to record observations with pencil and paper. However, keeping an observing logbook and drawing what you can see is an excellent way to train your eye to look for detail in astronomical objects, such as planetary surfaces and nebulae. And while digital cameras can produce stunning images, they are not always indicative of what can be seen with the eye.

There's nothing quite like a drawing for recording what you can see with your naked eye, or through the eyepiece of a telescope or binoculars. For deep sky objects, such as clusters, nebulae, and galaxies, the trick is to start with the stars. Focus on getting orientated by first drawing a few of the brighter stars visible through the eyepiece. Once they are in the right place, move on to filling in the fainter stars and sketching in the nebula or galaxy. To add different shades to nebulae or galaxies, simply vary the hardness of the pencil used, and soften edges by gently rubbing the paper with the tip of your finger. Brighter regions in a nebula can be highlighted with dabs of a clean eraser. With lunar sketching, try to place several prominent features accurately before filling in detailed lines, shading, and other features. Some lunar sketchers prefer not to "color in" the different shades of the lunar surface but instead use a pen to add different densities of dots that represent varying shades.

Many astronomers sketch in negative—using pencil on white paper—and then scan their sketches and use a computer graphics program to invert their images into bright stars on a black background. For solar observations, sketches made with yellow, red, and white pastels or chalk on black cardboard can be a fun and striking way to capture the Sun.

ROSETTE NEBULA
A sketch of a nebula, such as the Rosette Nebula (above), begins with the brighter stars (right), which establish a framework for dimmer objects. The brighter the star, the bigger its dot.

CRATER CLAVIUS
The key to sketching the surface of the Moon, such as the crater Clavius (above), is to concentrate on selected features that stand out. After lightly adding these shapes, use the angled side of a pencil or charcoal to add the shadowed areas, such as the crater sides and floors (right).

USING BINOCULARS

Binoculars are an astronomer's best friend because they are always ready for action and can provide outstanding views.

Why binoculars?

Small, portable, and affordable, binoculars require little or no setting up. The image they produce at the eyepiece is the "right" way up (in contrast to telescopes, which may produce an inverted image), making them simple and intuitive to aim and use. They provide a wide field of view, making them ideal for visual tours across large swathes of the sky and perfect for looking at larger objects, such as the Andromeda Galaxy (M31). Binoculars should certainly not be considered "second-best" to telescopes: in fact, most seasoned observers regularly use binoculars to supplement their telescopic astronomy.

Even a small pair of binoculars will reveal Jupiter's four largest moons; the prominent mountain ranges, craters, and smooth seas on the Moon; and the Milky Way's spiral arms, full of stars. Most galaxies and nebulae will appear as small gray misty patches, though some structure can be glimpsed in the brighter nebulae, such as M42 in Orion, in the clearest skies with moderate light pollution.

THE PLEIADES
Open star clusters such as the Pleiades appear in glittering, almost three-dimensional relief through binoculars.

OBJECTS TO SEEK OUT WITH BINOCULARS		
OBJECT NAME	**OBJECT TYPE**	**CONSTELLATION**
M42	Emission nebula	Orion
NGC 884 & 869	Open clusters	Perseus
M31	Galaxy	Andromeda
M22	Globular cluster	Sagittarius
M33	Galaxy	Triangulum
M13	Globular cluster	Hercules
M45	Open cluster	Taurus
NGC 5139	Globular cluster	Centaurus

Using binoculars

To get the most out of your binoculars, allow your eyes time to adjust to the darkness outside before starting your observations. Try reclining on a deck chair, lounger, or waterproof sheet on the ground; this will allow you to view the sky at higher angles without straining your neck. Alternatively, rest your elbows on the top bar of a fence or wall to provide support and stability.

First, seek out a bright star and use it to focus the binoculars (using the focus wheel) so that it appears as a bright, sharp point of light through the eyepieces. Methodically scan the area of sky around the star to become familiar with the greater detail now visible; once familiarized, you can start to navigate across the sky with the aid of a detailed star chart, hopping to celestial objects via different stars.

WIDE VISION
A pair of 10x magnification binoculars has a field of view of roughly 6–8°, making them great for looking at larger phenomena, such as large star fields that don't fit into a telescopic field of view.

RELAXED OBSERVING
Lying back on a deck chair lets you look up while bracing the binoculars against the arms.

ADJUSTING BINOCULARS

Pivot the optical tubes around the central bar to align both eyepieces with your eyes. Next, adjust the focus. Most binoculars have two focus adjustment wheels—one on the central bar, and another on one of the two eyepieces.

1 Close your eye on the side of the adjustable eyepiece. Focus on a distant object using the central focus wheel.

2 Close the other eye and focus on the same object using the eyepiece wheel.

Binoculars for astronomy

The job of binoculars is both to magnify objects and to collect more light than the naked eye, and so to make dim objects appear brighter. They contain glass lenses as well as prisms, which turn the image the "right" way around so they can be pointed easily at a target. Any binoculars you may already have for watching birds and wildlife can probably be used to observe the night sky. However, if buying a new pair specifically for astronomy, there are a few things to look out for.

Basic designs

Binoculars come in two basic designs, called roof- and Porro-prism. The Porro-prism design has the classic dog-leg shape and is wider and bulkier than the straight, more compact roof-prism. However, at the budget end of the market, Porro-prism binoculars tend to have better, brighter optics than roof-prism binoculars and represent better value for money.

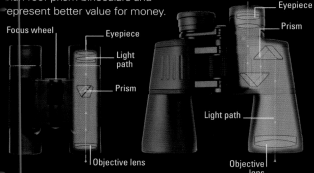

Focus wheel — **Eyepiece** — **Light path** — **Prism** — **Objective lens**

Eyepiece — **Prism** — **Light path** — **Objective lens**

ROOF-PRISM
In this design, light travels through a set of close prisms, allowing for a compact form.

PORRO-PRISM
A folded light path makes Porro-prism binoculars bulky; the objective lens is usually large, giving bright images.

Objective lens

Aperture measurement

OBJECTIVE DECISIONS
The size of the objective lens, or aperture, of a pair of binoculars determines their light-gathering ability. Larger apertures usually produce brighter images.

Magnification and aperture

Binoculars are usually labeled with two numbers—for example, 10x50, pronounced "ten by fifty." The first of the two numbers is the magnifying power of the binoculars and indicates how many times they enlarge the naked-eye view of an object. High magnifications may seem attractive, because they make objects appear larger, but they also amplify every jolt as you try to hold the binoculars steady, making the image jump around in the eyepiece. Ten times magnification is a good compromise.

The second number is the diameter of the front lens (the objective lens) in millimeters. It is this lens that gathers the light from dim objects, so bigger—up to a point—is better. Ideally, you should look for a pair of binoculars with an objective lens of at least 50mm; larger objective lenses will make the binoculars bigger, heavier, more expensive, and harder to hold still for long periods.

Larger models, with magnifications of as much as 25x and objective lenses 120mm wide, provide exceptional views of the sky, though they are costly and need to be mounted on a sturdy tripod with an adapter to achieve shake-free views.

GLASS AND OPTICS

In general, the more you pay for your binoculars, the better their optical quality. Look for models that use high-grade Bak4 glass and have "multi-coated" or "fully multi-coated" lenses. These coatings reduce the amount of light scattered by the optics and so improve image contrast. Another initial consideration when choosing binoculars is the "exit pupil." This provides an indication of how bright the view is through the binoculars. It is measured in millimeters and is calculated by dividing the diameter of the objective lenses by their magnification. Binoculars with an exit pupil of between 5 and 7mm are ideal.

Coating on lenses

LARGE BINOCULARS
High-magnification, large-aperture binoculars need to be mounted on a tripod. Inexpensive tripod adapters secure them in place.

TELESCOPE OPTICS

A telescope opens up the night skies to new levels of observation. Choosing a telescope that meets your needs requires some understanding of basic optical concepts.

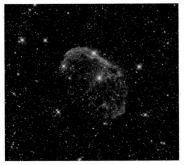

THE RIGHT TELESCOPE FOR THE TASK
Telescopes that get the best views of deep-sky objects are not necessarily useful for observing closer objects, such as the Moon.

The parts of a telescope

The function of any telescope is simple—to collect light and magnify a view. All telescopes consist of three basic parts: a tube assembly that contains the telescope's optics, which may be lenses, mirrors, or both; a mount that supports the tube and allows the telescope to be pointed precisely; and a tripod (or pier) that provides a solid base onto which the tube and mount can be secured.

Telescopes, mounts, and tripods come in a huge variety of designs. When buying your first telescope, your choice will depend on budget, portability, and, importantly, your astronomical interests. Some telescopes provide a good compromise, allowing clear views of diverse objects, but such jacks-of-all-trades are usually masters of none. The abilities and applications of each telescope depend on its optical specifications, which are explored below.

THE SUM OF ITS PARTS
The optics of the telescope, the accessories fitted, and the type of mount used will all affect overall performance.

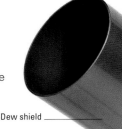

Dew shield

Light-gathering ability

It is perhaps surprising to learn that most objects in the night sky are hard to see not because they are too small, but because they are too faint. It follows that the most important measure of a telescope is not its raw magnifying power but the size (aperture) of its main lens or mirror—or "objective"—which is usually measured in millimeters or inches. The larger the main lens or mirror, the more light it will be able to collect, and the more detail the telescope will show. Compare your own in-built telescope—your eye, which has an aperture (the pupil) of around 7mm—with a moderately-priced telescope with a main lens 60mm wide: the telescope harvests a staggering 70 times as much light as your naked eye, and so makes faint objects appear bright enough to see.

120mm APERTURE

66mm APERTURE

OBJECTIVE SIZES
Aperture, the diameter of the objective lens, is the most important specification of a telescope.

LONG FOCAL LENGTH

SHORT FOCAL LENGTH

FOCAL LENGTH
For a given eyepiece, a long-focal-length telescope will produce a more magnified view than a telescope with a shorter focal length.

Focal length and ratio

After aperture, the most important specification of a telescope is its focal length, which is the distance from the objective lens to the focal point (see right): for a given eyepiece, a long-focal-length telescope will produce a more magnified image, with a smaller field of view, than a shorter-focal-length telescope. Long-focal-length telescopes tend to be favored for observing bright objects, such as the Moon, planets, and double stars. Short-focal-length telescopes, on the other hand, are popular for observing targets such as the rich star fields of the Milky Way.

Dividing the focal length (in millimeters) by the telescope's aperture (in millimeters) gives its focal ratio or "f-number." This is mainly of interest to astrophotographers for whom "fast" telescopes with a lower focal ratio, typically around f/5, are used for imaging faint deep-sky objects.

Optics and magnification

The optical heart of a telescope consists of two components, the tube and the eyepiece. The tube, blackened on the inside to prevent reflection, has at one end a main lens or curved mirror, called the objective. This focuses light from a distant object, such as a star, on to one point—called the focal point—located just in front of the eyepiece at the other end. If you were to place a sheet of tracing paper at the focal point, a crisp image of the star would be projected onto the sheet. The eyepiece is in essence no more than a magnifying glass, allowing the image at the focal point to be magnified to different degrees, depending on the eyepiece selected.

The magnifying power of a telescope is a ratio, which describes how big an object appears in the eyepiece compared to a naked-eye view—25x, 100x, and so on—and it depends on the focal lengths of both the telescope and the eyepiece. To calculate the magnification of a setup, simply divide the focal length of the telescope by that of the eyepiece. So a telescope with a focal length of 2,500mm used with an eyepiece with a focal length of 25mm gives a magnification of 100 times.

Changing the eyepiece changes magnification, but only within practical limits set by the aperture of the telescope (see table, right). As a general rule, it is wise not to use an eyepiece that gives a magnification greater than twice the size of the telescope's objective (in mm).

Objective lens

Aperture

Tube

Focal length

Mount

Tripod head

Finderscope

Focal point

Eyepiece

BASIC OPTICS
This telescope is a refractor, using lenses (not mirrors) to form an image. It has an aperture of 120mm and a focal length of 1,000mm giving it a focal ratio of f/8.3.

APERTURE AND MAGNIFICATION

The maximum magnification of a telescope depends on its aperture. Magnifications of 300x or more are of little use, as they only magnify atmospheric turbulence.

TELESCOPE APERTURE		HIGHEST USABLE MAGNIFICATION	FAINTEST OBJECTS VISIBLE	RESOLVING POWER
mm	inches			
60	2.4	120x	mag. 11.6	1.9"
80	3.1	160x	mag. 12.2	1.5"
90	3.5	180x	mag. 12.5	1.3"
100	3.9	200x	mag. 12.7	1.2"
120	4.7	240x	mag. 13.1	1.0"
150	5.9	(300)x	mag. 13.6	0.8"
200	7.9	(400)x	mag. 14.2	0.6"
250	9.8	(500)x	mag. 14.7	0.5"

9mm EYEPIECE

25mm EYEPIECE

EYEPIECE MAGNIFICATION
Switching eyepieces changes the magnification of a particular telescope. High-magnification eyepieces are not always the best because they tend to reduce image contrast and brightness, and provide a very narrow view of an object.

Eyepieces

A telescope eyepiece consists of several lenses in a metal cylinder, which usually has a number—the focal length, in millimeters—printed on its side. A small range of eyepieces will provide a good range of magnifications. One with a short focal length of 9mm is ideal for high-magnification observing of the Moon, planets, and double stars. A 25mm focal length is good for general use, but for wide-field observing of open star clusters and nebulae, an eyepiece with a longer focal length of around 40mm is perfect. Another useful eyepiece accessory is the Barlow lens. This lens slots in front of the eyepiece, increasing its magnification by a given number of times. This number is printed on the lens itself and is usually either 2x or 3x.

RESOLVING POWER

The resolution of a telescope is a measure of how much detail it can "see"—for example, whether it can distinguish between the two stars of a binary star. It is measured in tiny angles of separation (arcseconds). In theory, the resolution is set by the telescope's aperture, but in practice, optical defects and atmospheric turbulence set the limits.

40mm **25mm** **9mm** **2X BARLOW LENS**

TELESCOPES AND MOUNTS

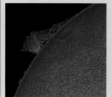

There are many designs of telescopes and mounts available today which vary greatly in their capabilities, cost, and ease of use.

Lenses and mirrors

All telescopes are designed to brighten and magnify views of celestial objects. The three main designs of telescope—reflector, refractor, and catadioptric—achieve this in slightly different ways and have different advantages and drawbacks. The reflector uses a series of mirrors to collect light from a distant object and focus it on an eyepiece, while the refractor uses a lens (or series of lenses) to do this. A catadioptric is a hybrid of the two designs that uses both lenses and mirrors to collect and focus light.

SPECIALIST SCOPES

Once the province of professional astronomers, dedicated solar telescopes are now affordable to amateurs. These instruments, which transmit only a narrow-wavelength band of light, open up the Sun to observation.

DYNAMIC PHENOMENA
Hydrogen-alpha solar telescopes reveal spectacular plasma clouds (prominences) on the Sun's limb.

OPTICAL DESIGNS

TYPE	DESCRIPTION	ADVANTAGES	DISADVANTAGES
REFRACTOR	The refractor is the "classic" long-tube astronomical telescope and it is the direct descendent of the telescopes built by Galileo. In its simplest form it consists of a tube with a large lens (the objective) at one end that collects light and focuses it down the tube where a smaller lens, the eyepiece, magnifies the image and throws it into the viewer's eye. Refractors are ideal for looking at objects such as the Moon, planets, and double stars.	Robust and requires little maintenance, making it suitable for children and for carrying around. Provides sharp images with good definition. Has no mirrors or obstructions within the tube, giving high-contrast images.	Generally more expensive than reflectors of a similar aperture. Prone to "chromatic aberration"—an imperfection in which different colors are focused on the lens at different points, giving stars a faint halo of color. Can be awkward when observing objects high in the sky.
REFLECTOR	The Newtonian reflector—invented by the English scientist Sir Isaac Newton—consists of an open-ended metal, plastic, or wooden tube with a (usually) parabolic mirror at its base. This mirror collects the light and reflects it back up the tube where it meets another, smaller mirror suspended on an assembly of thin vanes called a "spider" within the tube; the light is bounced out at right-angles to the tube, where it can be magnified by the eyepiece.	Relatively inexpensive, providing lots of aperture for the money—ideal for deep-sky observing. Location of the eyepiece gives comfortable viewing. Free from chromatic aberration (see above) but prone to "coma"—a flaw that makes stars at the edge of the field look wedge-shaped.	Relatively fragile; needs regular maintenance to keep the mirrors properly aligned. The mirror coatings also degrade after several years. Optically compromised by the presence of the secondary mirror, which obstructs light and can produce "spikes" that seem to emanate from bright objects.
CATADIOPTRIC	This modern design uses both lenses and mirrors to form an image. It consists of a short tube, the bottom of which holds the primary mirror. At the top end of the tube a thin, full-diameter lens (or corrector plate) compensates for optical imperfections. On the inside of the corrector plate is a small secondary mirror, which reflects light collected by the primary mirror through a hole in the primary mirror toward the eyepiece mounted on the end of the tube.	Compact, portable design, with long focal length in a short tube. Can be mounted on a convenient fork mount (see opposite, top right). Far less prone to the refractor's "chromatic aberration" and the reflector's "coma." Good general-purpose instrument.	Tends to be more expensive than a reflector of equivalent aperture. Complex design means some light is lost before it reaches the eye, making the image dimmer than it would be from a same-size reflector—though special coatings do help reduce this light loss.

Telescope mounts

The job of the telescope mount is to provide a firm, stable platform for the optics: the mount should support the telescope without the instrument wobbling and also allow it to be aimed accurately toward a target in the sky. The mount is a key component of your setup, especially if your interests lie in astrophotography: when choosing your equipment, bear in mind that even the best telescope will suffer on a poor mount. When buying, give as much consideration to the quality of the mount as to the quality of the telescope itself.

DOBSONIAN   **FORK MOUNT**

PORTABLE TRACKING MOUNTS
Compact, portable tracking mounts can carry small telescopes and DSLR cameras. They are useful for astrophotography, for observing from relatively inaccessible dark sky locations, and for traveling.

DESIGN VARIANTS
The Dobsonian and fork mounts are variants of the alt-az design (see below). Dobsonians are easy to set up and good for deep-sky observing. Fork mounts are often sold with go-to type telescopes (see p.62).

MOTORIZED TRACKING

A small electric motor drive attached to the right ascension (RA) axis of a polar-aligned equatorial telescope will make it track an object as it appears to move through the sky (though you still need to locate the object in your eyepiece in the first place). The accuracy of this tracking depends on the precision of the mount's mechanism.

FINE-TUNING
Small adjustments in position can be made through a control unit attached to the motor.

Altitude-azimuth (alt-az) or equatorial?

Telescope mounts come in two main varieties, the altitude-azimuth, or alt-az, mount, and the equatorial. The simpler alt-az mount moves in much the same way as the head of a tripod, allowing movement of the telescope around two axes—up and down through 90° (altitude) and left or right through 360° (azimuth). It is easy to point and ideal for casual stargazing. The problem is that stars do not move (or appear to move) up and down and left to right—their apparent motion is in a circle around the celestial pole, which makes tracking stars difficult with an alt-az mount.

This is where the equatorial mount comes into its own. Like the alt-az, it also has two axes of rotation at right angles to one another: they are called right ascension (RA) and declination (DEC). However, once the telescope is aligned with the celestial pole (see p.60), you need only move it around one axis (RA) to follow the path of an object in the sky. The ability to track an object smoothly, rather than in a series of up-down/left-right steps, makes the equatorial mount suitable for astrophotography, where a single exposure may last many minutes, and makes it far easier to keep an object in the eyepiece. Furthermore, the equatorial mount allows you to find stars if you know their celestial coordinates, using the setting circles (see p.61) and can readily be motorized. Using an equatorial mount will teach you all about the details of the stars' apparent motion and celestial coordinates; but if you just want to get out there and start observing, consider buying a go-to telescope (see p.62), which does all the hard work for you.

ALT-AZ MOUNT
Alt-az mounts are usually lighter and more compact than equatorial mounts. Tracking a celestial object requires moving the telescope in two directions at the same time: some people find this easy, while others struggle.

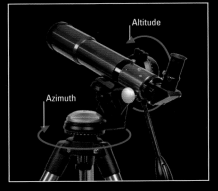

Altitude

Azimuth

TRACKING STARS
Stars appear to move in circular paths around the celestial pole. The motion of the alt-az mount is stepped, while the nearly circular movement of the equatorial mount allows stars to be tracked easily.

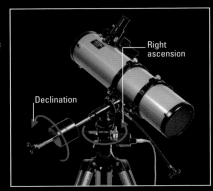

Right ascension

Declination

Finding your target

Accurately aiming a telescope with a narrow field of view at a faint star can be difficult, and even large targets can be hard to locate without the help of a finderscope. This is a small refracting telescope that sits on the side of the main telescope, enabling precise alignment of the main scope. A finderscope has a far smaller magnification (typically 6x or 8x) than the main telescope and provides a wide, bright-field image of the sky; once its crosshairs are centered on a night-sky object, that object will also be visible through the main telescope. The finderscope must be aligned accurately with the main telescope (see below) each time the pair are used.

Some astronomers prefer to use a type of finder that does not magnify the view; this is called a "reflex" or simply "red dot" finder. This type of finder has a red dot or target pattern projected onto a piece of glass or plastic, making the dot or pattern appear to float in the field of view. Once the target is centered on the object, the object will also appear in the telescope. Red dot finders are easier to use than finderscopes because the whole sky remains in view, and objects are seen in context. Objects are also easy to align because the view is not reversed (as is the case with a finderscope).

FINDERSCOPE

A finderscope magnifies the night sky and gives a field of view of around 5–8°. A crosshair helps to center the target in the finderscope. The image through the finderscope is inverted, which can, at first, make finding objects frustrating. Most entry-level telescopes come with a basic finder, but it may be worth upgrading as you progress.

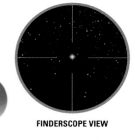

FINDERSCOPE VIEW

Eyepiece — — Mounting bracket

RED DOT FINDER

A red dot finder indicates where the telescope is pointing by projecting a small red dot onto a piece of transparent glass or plastic. The wider sky remains visible, making it intuitive to use. The brightness of the dot can sometimes be adjusted with a built-in dimmer switch.

Sight —

RED DOT FINDER VIEW

Alignment adjustment wheel — — Mounting bracket

Aligning your finderscope

Periodically aligning your finderscope or red dot finder with your telescope is a simple but essential task that makes observing so much easier. Carry out a rough alignment during daylight hours using an Earth-based object as a target. This has the great advantage that it will not move during alignment, making the job much easier. Remember never to point your finderscope or telescope at the Sun when performing alignment.

1 AIM THE TELESCOPE
During daylight hours, choose a distant object, such as a lamppost, and align your main telescope carefully so that the target sits right in the middle of your field of view.

2 CHECK THE FINDERSCOPE
Make sure the barrel of your finderscope or red dot finder is roughly parallel with your telescope. Look through its eyepiece; the target will probably be off-center.

3 ADJUST THE FINDERSCOPE
Use the adjustment screws or wheels on the finderscope to make small changes in the alignment of the telescope. Keep checking the alignment of the two instruments.

4 FINALIZE ALIGNMENT
Once the two scopes are correctly aligned, tighten the adjustment screws. Aligning your main telescope with the finderscope every now and then will make using it much easier.

Altered views

The view through some astronomical telescopes is inverted—up is down, left is right—and you see the sky as if standing on your head. This can make life difficult when searching for objects in the sky or comparing a telescope view to a star chart in a book. Many astronomers use accessories called diagonals, both to correct for this inversion and to provide a more comfortable viewing position. The star diagonal uses a mirror or a prism to turn the light from the telescope through 90°—turning the light through a right-angle means that you can look down through the eyepiece instead of up, which lessens the strain on your neck. The star diagonal also corrects the image vertically (so up is up and down is down), but flips it horizontally. A different type of diagonal, called a "correct image" or "erect image" diagonal, is needed to achieve a correct image; this rights the image vertically and horizontally.

This increased comfort and ease comes at a cost to the image, because the optical assembly of the diagonal can absorb and disperse up to 10 percent of the light entering the telescope. For this reason, many astronomers prefer to become accustomed to the initially strange movements of the image. You can practice this in daylight hours by pointing your telescope at a lamppost in the distance during the day and experimenting with movements of the telescope.

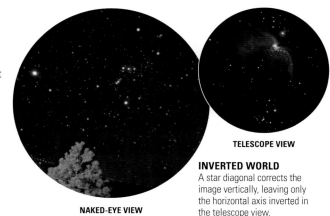

TELESCOPE VIEW

NAKED-EYE VIEW

INVERTED WORLD
A star diagonal corrects the image vertically, leaving only the horizontal axis inverted in the telescope view.

DIAGONALS
A diagonal fits easily between the telescope and the eyepiece. One of its main advantages is providing a comfortable viewing position with reflectors and catadioptrics, where the eyepiece is at the back of the telescope.

Using filters

Filters are pieces of optical glass that block out specific parts of the color spectrum, while allowing other wavelengths (colors) to pass, and they can significantly enhance views of the night sky. There are many hundreds of types, some with highly specialized functions, but the simplest type is a colored filter used to enhance the details of planetary features. A yellow filter, for example, can help bring out more detail when viewing Mars's polar ice caps. A blue filter will greatly increase contrast in the belts and Great Red Spot of Jupiter. There are Moon filters that reduce the intensity of light when looking at the very bright surface of the Moon; filters that subdue the orange glow of light pollution; and highly specialized narrow band filters that allow only specific wavelengths—associated with the light emission of different glowing gases—to pass. Other specialized filters are used when taking pictures of everything from galaxies to planets.

DEEP SKY FILTERS
Deep sky observers use specialized narrow-band filters to create more contrast between the background blackness of space and nebulae they are trying to observe—here the Veil Nebula, a large supernova remnant in the constellation Cygnus.

FITTING A FILTER
Filters typically screw into the rear of the eyepiece, which is then attached to the telescope as usual. Filters may enhance a particular view, but they will also cut down the amount of light reaching the eye.

BUYING AND SETTING UP

Choosing a telescope to suit your needs, and setting it up carefully and accurately, will greatly enhance your observing experience.

Balancing and aligning

After buying a telescope, it is important to take enough time to set up its optics, tripod, and mount properly. Careful setup will leave you with a well-aligned and balanced telescope that is a joy to use, and that will require minimum tweaking during those precious observing hours. Each telescope is different, so be sure to read the instructions provided before starting, or better still, ask an experienced astronomer to take you through the basics. Below is a brief and general guide to the main points of setting up a typical amateur telescope—a reflector on a motorized equatorial mount. You will probably leave your telescope partly set up between observing sessions, so some of the steps will only need to be carried out once.

BUYING ADVICE

Before choosing your equipment, think carefully about what type of observing you want to do and take advice from a specialist optical store. Set a clear budget and—as a general rule—go for the biggest aperture you can afford: ideally, at least a 100mm reflector/catadioptric or a 60mm refractor. Don't skimp on the mount and tripod—they are as important as the optics of the telescope. You will learn a lot about celestial coordinates if you buy and use a basic equatorial mount, but if you want to observe with more ease and speed, consider a motorized go-to telescope (see pp.62–63).

CHILD'S REFLECTOR

GO-TO TELESCOPE

Setting up a telescope and mount

1 LEVEL TRIPOD
Set up your tripod on solid, level ground. Use a level to check that the top plate of the tripod is horizontal and adjust the tripod legs accordingly.

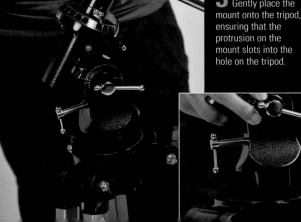

BALANCED TRIPOD

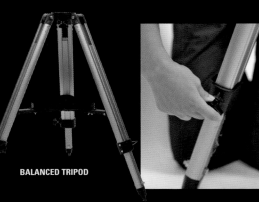

2 ADJUST LEGS
Avoid extending the sections of the tripod legs to their full extent, because this tends to make the platform less stable and gives you no latitude for fine adjustment of height later. Double-check that the locks on the legs are secure.

3 PLACE MOUNT
Gently place the mount onto the tripod, ensuring that the protrusion on the mount slots into the hole on the tripod.

4 SECURE MOUNT
Tighten the mounting screw from beneath the tripod head, making sure it is secure.

5 ATTACH MOTOR DRIVE
Attach the motor drive to the mount and ensure that the gears of the motor are correctly engaged with those on the mount.

**TRIPOD COMPLETE
WITH MOUNT**

Circular tube
mounting rings

6 ALIGN NORTH

When using an equatorial mount, now is the time to check that the right ascension axis (the long part of the central "T" of the mount) is pointing roughly toward the north or south) celestial pole (see p.60 for more detail).

7 ADD COUNTERWEIGHTS

Slot the counterweights onto the counterweight shaft and use the nut to secure the weights in position. There is usually a safety screw at the end of the shaft which functions to stop the counterweights from sliding off, should the main nut fail. Be sure to replace this safety screw after positioning the weights.

8 MOUNT THE TELESCOPE

With the mount on the tripod, you can mount the telescope tube. Place the telescope tube inside the pair of circular mounting rings and clamp the rings tight with the screws around the tube.

10 ADD THE MOTOR UNIT

Plug the drive controller into the motor unit, but do not connect it to the power supply.

9 ADD FINE ADJUSTMENT CABLES

Screw in the fine adjustment cables: these allow you to make small changes to right ascension and declination when observing.

11 FIT THE FINDER AND EYEPIECE

Fit the desired eyepiece and finderscope or red dot finder onto the telescope. Secure them in position with the tightening nuts. Align your finder with the main telescope (see p.56).

BALANCING THE MOUNT

The next important step is to balance the mount. This will allow your telescope to move smoothly, and hold its desired position when you do move it. The mount needs to be balanced in both its axes of movement, starting with the right ascension (RA).

12 BALANCE RIGHT ASCENSION (RA)

Support the telescope with one hand and loosen the clutch that stops the mount from moving on the RA axis. Undo the nut securing the counterweight and slide it up and down the counterweight shaft until it balances the weight of the telescope. Secure the nut at this point.

13 BALANCE DECLINATION (DEC)

To balance the mount in the other dimension of movement—declination— position the mount so that the telescope is out to one side, then loosen the clutch that allows motion on the declination axis. Be sure to support the telescope with one hand, as it will most likely tilt to one side. Loosen the mounting rings holding the telescope tube in the mount, and slowly slide the tube to and fro until it is balanced. Tighten the clutch to secure the mount.

14 POWER UP

Connect the motor to the power supply and then fine-tune the polar alignment of the telescope (see p.60).

**FULLY ASSEMBLED
TELESCOPE WITH
TRIPOD AND
MOUNT**

Polar alignment

The night sky revolves around the rotational axis of the Earth, at either end of which are the celestial poles. The beauty of an equatorial telescope mount is that, once correctly aligned with this axis of rotation, it can track the stars as they appear to move. Astronomers can also use the equatorial mount to find objects in the sky if their celestial coordinates are known. Aligning the polar axis of the mount is often a cause for concern for beginners, but with practice can become easy. Remember that polar alignment involves aligning the mount rather than the telescope.

CIRCULAR MOTION
The stars appear to move in circular paths around the celestial pole.

NORTH CELESTIAL POLE   **SOUTH CELESTIAL POLE**

Where are the celestial poles?

The position of the north celestial pole is very close to the star Polaris (in Ursa Minor). For casual observing, it is sufficient to align an equatorial mount with this star, but for more accurate observing, you will need to use a polarscope (see below) to locate the north celestial pole itself.

The position of the south celestial pole is harder to find, because there are few prominent stars nearby. To locate the south celestial pole, draw two imaginary lines across the sky. The first is drawn perpendicular to a line that joins the stars Alpha and Beta Centauri, while the second travels down the long axis of the Southern Cross. The south celestial pole lies where these two lines meet. A polarscope makes accurate alignment much easier.

Estimating polar alignment

For casual observing, you can carry out an estimated polar alignment very quickly. To do this, use a normal compass to find north (if you are observing in the northern hemisphere) or south (if you are observing in the southern hemisphere). Lift up the telescope, mount, and tripod, and point the polar axis of the mount to the north (or south). Then, use the adjustment knob at the bottom of the mount to set your latitude as accurately as possible on the scale. You can find your latitude by consulting a map or by searching the Internet. The mount is now aligned approximately with the pole.

The polar (or right ascension) axis of the mount is the long part of the "T" that forms the center of the mount, and is at a right-angle to the counterweight shaft.

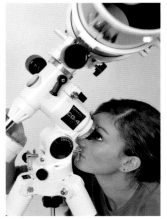

POLAR SIGHTING
Looking through a northern hemisphere polarscope (left), you will probably see a reticule engraved with several constellations and a circle offset from a central cross hair. Locate the star Polaris and move the axes so that it sits in the circle in the correct orientation with respect to the constellations (below).

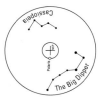

QUICK ALIGNING
Point the polar axis of the equatorial mount to the north (left). Loosen the nut on the mount to set your observing latitude on the scale at the base of the mount (above).

Accurate alignment

For longer periods of observing, or to accurately track objects for astrophotography, you need to align the mount precisely on the north (or south) celestial pole. The best way to do this is to use a mount equipped with a polarscope. This is, in effect, a small telescope set within the polar axis of the mount. Looking through the polarscope will reveal a series of markings that show the position of the celestial pole in relation to nearby stars, such as Polaris in the northern hemisphere. Small adjustments to the altitude and azimuth adjustment knobs of the polar axis allow the view through the telescope to be matched up with the markings, so that the mount is aligned to the celestial pole.

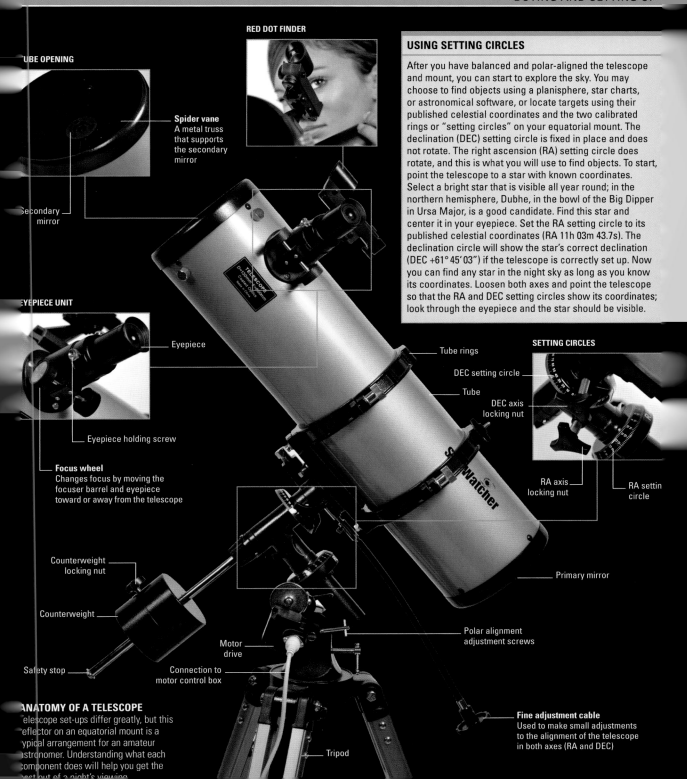

TUBE OPENING

Spider vane
A metal truss that supports the secondary mirror

Secondary mirror

RED DOT FINDER

USING SETTING CIRCLES

After you have balanced and polar-aligned the telescope and mount, you can start to explore the sky. You may choose to find objects using a planisphere, star charts, or astronomical software, or locate targets using their published celestial coordinates and the two calibrated rings or "setting circles" on your equatorial mount. The declination (DEC) setting circle is fixed in place and does not rotate. The right ascension (RA) setting circle does rotate, and this is what you will use to find objects. To start, point the telescope to a star with known coordinates. Select a bright star that is visible all year round; in the northern hemisphere, Dubhe, in the bowl of the Big Dipper in Ursa Major, is a good candidate. Find this star and center it in your eyepiece. Set the RA setting circle to its published celestial coordinates (RA 11h 03m 43.7s). The declination circle will show the star's correct declination (DEC +61° 45′ 03″) if the telescope is correctly set up. Now you can find any star in the night sky as long as you know its coordinates. Loosen both axes and point the telescope so that the RA and DEC setting circles show its coordinates; look through the eyepiece and the star should be visible.

EYEPIECE UNIT

Eyepiece

Eyepiece holding screw

Focus wheel
Changes focus by moving the focuser barrel and eyepiece toward or away from the telescope

Tube rings

DEC setting circle

Tube

DEC axis locking nut

SETTING CIRCLES

RA axis locking nut

RA setting circle

Counterweight locking nut

Counterweight

Safety stop

Motor drive

Connection to motor control box

Primary mirror

Polar alignment adjustment screws

ANATOMY OF A TELESCOPE
Telescope set-ups differ greatly, but this reflector on an equatorial mount is a typical arrangement for an amateur astronomer. Understanding what each component does will help you get the best out of a night's viewing.

Tripod

Fine adjustment cable
Used to make small adjustments to the alignment of the telescope in both axes (RA and DEC)

GO-TO TELESCOPES

Computer technology, paired with affordable optics, has given us the go-to telescope, an instrument that makes navigating the night sky easy, even for novices.

Quick and easy observing

Many amateur telescopes are sold with mounts equipped with electric drives and control handsets (see p.55) that allow the instrument to track stars across the night sky. The go-to is a type of telescope that is equipped with motors, but which takes automated observing further still. Once aligned to the night sky, a go-to telescope can automatically locate and move to a multitude of different celestial objects, without assistance from the observer. It has a computer handset (called a paddle) on which the celestial destination is selected from a "menu" of targets: containing sophisticated electronics, the paddle communicates with the mount, telling it where to go and how to get there. And for the astronomer who is unsure what to look for during a night's observing, some go-to handsets can even produce tours of the evening's most notable celestial objects.

SELF-GUIDED TELESCOPES
A go-to telescope will not improve your view of celestial objects, but it will make them easier to locate. Using a go-to telescope requires you to have very little knowledge of celestial coordinates or even of navigating by the constellations.

How does the go-to work?

For a go-to telescope to work, it must be aligned to the night sky. For equatorial mounts, this may mean carrying out polar alignment; fork-mounted go-to telecopes will need to be moved into a predetermined "home" position, specified in the manual. Once the mount and handset electronics are switched on, the date, time, and location need to be entered. Controlled by its onboard computer, the telescope will then move to where it calculates a bright star should be at that time and date. You then confirm that the star is indeed in the eyepiece or make minor adjustments with the handset to center the star. This alignment process is repeated for up to three widely separated stars, after which the computer has an accurate picture of its location under the celestial sphere. From then on, the go-to telescope can move automatically to thousands of objects in its database, because it "knows" exactly where it is pointing.

COMPUTER CONTROL
Some go-to telescopes can be hooked up to a computer and controlled by planetarium software packages; some paddles can be updated via a computer and the Internet so that newly discovered objects can be found and tracked.

Setting up a typical go-to telescope

1 MOUNT THE TELESCOPE
Use a level to check that the tripod is fully level. Gently lower the mount and telescope onto the tripod head and secure it in position.

2 SET POSITION
Move the telescope tube and mount into its home position. For a fork-mounted go-to telescope (shown), this may just mean aligning two arrows; but an equatorial mount will need polar alignment (see p.60).

3 PREPARE THE TELESCOPE
Connect the mount to the power pack and switch on the mount. Remove the lens cover

4 ENTER START DATA
Enter the date, time, and location into the handset as prompted. On some go-to telescopes, you select your location from a menu

GPS TELESCOPES

Some top of the range go-to telescopes take automated observing even further. Equipped with GPS (Global Positioning System) receivers, they automatically set the date, time, and position every time the instrument is switched on. Some even have sensors that detect if the telescope is on an uneven base, and will automatically make sure it is levelled and pointing the correct way before alignment begins.

PLANET TRACKER
Go-to scopes will find and track stars, deep-sky objects, and planets, such as Jupiter. Unsurprisingly, good go-to scopes are more costly than conventional scopes, but are well-suited to the cash-rich, time-poor astronomer.

5 ALIGN THE TELESCOPE
Alignment methods vary between models. Typically, you select a bright star, such as Sirius, to align with. Select the star on the handset, and the telescope will move automatically to where it calculates the star should be.

6 ADJUST ALIGNMENT
Look through the telescope's eyepiece; the selected star should be visible. If it is not, use the finderscope or red dot finder to center it in the main eyepiece using the directional buttons on the handset. Repeat steps 5 and 6 to align to a further two or three stars.

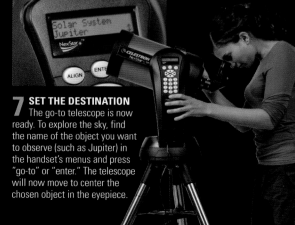

7 SET THE DESTINATION
The go-to telescope is now ready. To explore the sky, find the name of the object you want to observe (such as Jupiter) in the handset's menus and press "go-to" or "enter." The telescope will now move to center the chosen object in the eyepiece.

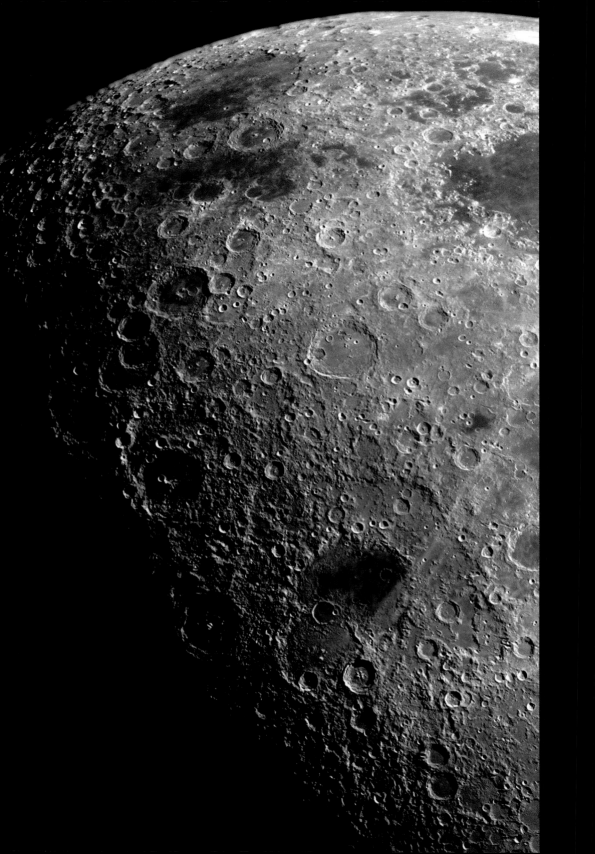

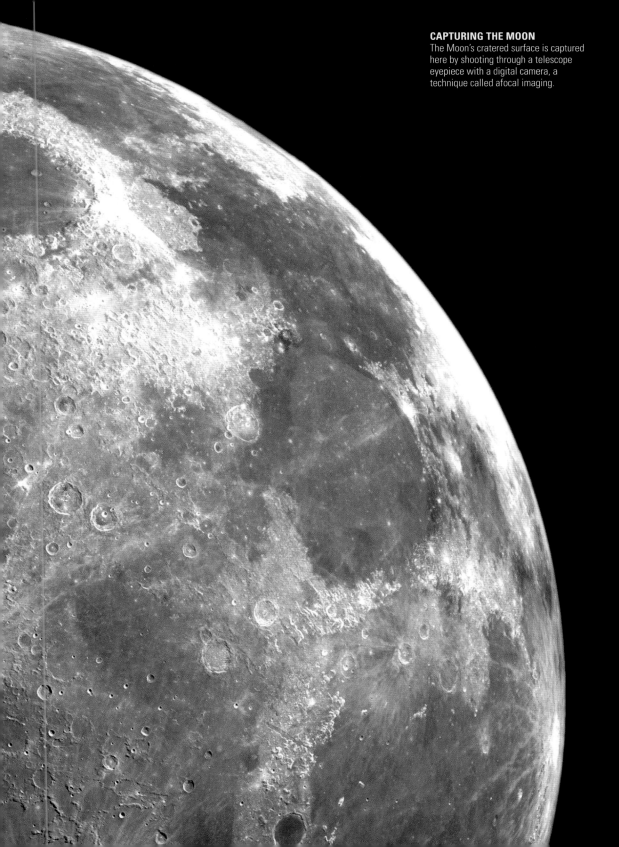

ASTROIMAGING

With a Universe of subjects to choose from and a constantly changing sky, astroimaging can be a hugely rewarding and exciting experience for astronomers at all levels.

Imaging choices

There are many ways to take pictures of the night sky: the simplest methods require no more than a basic "point-and-shoot" digital camera and a telescope, while the most advanced demand precisely driven telescopes, dedicated astronomical imaging devices, and specialized image-processing software. Like any skill, astroimaging requires patience and a willingness to learn, but with an understanding of the basic principles anyone can start to take great pictures.

MARS APPROACHES
Mars's polar cap is visible in a webcam image, as shown here during the planet's 2007 opposition, when Mars was close to the Earth in its orbit.

Afocal imaging

One of the simplest and least expensive kinds of astrophotography is called afocal imaging, or digiscoping. At its most basic, this involves aligning the lens of a point-and-shoot digital camera or smartphone with the eyepiece of a telescope, or even binoculars, and taking a picture. The attraction of afocal imaging is its ease and immediacy. Most point-and-shoot cameras and smartphones will be able to take reasonable pictures of the Moon and bright planets. When choosing a camera specifically for digiscoping, a large live-view LCD screen and the ability to vary the exposure length and ISO level manually are useful features to look out for.

STABLE MOUNT
A simple and inexpensive adapter allows the lens of a smartphone to be aligned accurately and securely with the eyepiece of a telescope or spotting scope.

AFOCAL IMAGING TIPS

- Align the camera lens as precisely as possible with the eyepiece of the telescope.

- Experiment with the distance between the camera lens and the telescope eyepiece to achieve optimum results.

- Mount the camera on a sturdy tripod while pointing it down the telescope eyepiece, or use a dedicated adapter (above) to secure the camera to the eyepiece.

- Switch off the camera flash and try varying the camera's exposure away from auto if your camera allows—trial and error will guide you to the best settings.

- Use the camera's self-timer to take photos: this will minimize any camera shake when you press the shutter button.

MOON MOSAIC
This picture of the Moon's surface shows what can be achieved with a simple digiscope. The image, which clearly shows the Moon's cratered surface and smooth dark seas, is made up of several digiscope images pieced together in a free graphics program.

Webcam astronomy

One big problem for astronomers taking pictures is the distorting undulation of the Earth's atmosphere, which causes "seeing" or blurring of objects (see p.45). Single images taken when the "seeing" is momentarily poor can be blurry and lack fine detail. An ingenious way around this problem is to take many individual images every second; in other words, a video. A short video can then be broken down with the help of special software—such as the excellent Registax (see below right) and AutoStakkert!—into its constituent frames. The best frames, which show the sharpest views, least distorted by the undulating atmosphere, are then filtered out by the software and stacked together to create a final image that shows much more detail and clarity than the raw video.

Remarkably, some webcams are ideal for taking videos of the Moon and planets, and many amateur astronomers have used them with their telescopes to take excellent pictures. Today, dedicated lunar and planetary cameras, similar to webcams, can be bought and used to capture very detailed images.

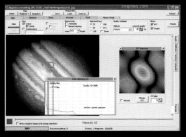

HIGH-FRAME-RATE IMAGING
Dedicated astronomy cameras simply slot into the telescope in place of the eyepiece (above). Some have dedicated capture software, while video processing can be done with programs such as Registax (right).

Using digital single-lens reflex cameras

The growing availability of digital cameras and the inventiveness of astroimagers means that it is possible to take stunningly detailed images of the night sky, from wide-field pictures of constellations to colorful images of faint nebulae and galaxies.

Digital single-lens reflex cameras (DSLRs) are not dedicated astronomical cameras but have proven themselves to be great for astroimaging in recent years. DSLRs have many advantages over their smaller point-and-shoot relatives. The shutter of a DSLR can be left open for long periods of time, which allows the light-sensitive sensor in the camera to gather as much light from faint objects as possible—this is very important when imaging galaxies or nebulae. DSLR sensors are also larger than those in point-and-shoot cameras, which means that they can register lower levels of light and are less prone to electronic "noise" (unwanted interference that registers on the sensor and appears on the image as fine speckles) so producing a cleaner image.

WIDE-FIELD IMAGING
A short exposure of about 30–60 seconds using a DSLR mounted on a tripod will show the constellations, the star fields of the Milky Way, and perhaps some brighter nebulae.

Versatile cameras

DSLRs are extremely versatile because they have interchangeable lenses. A standard zoom lens, which usually comes with the camera and typically has a focal length of around 18–55mm, is ideal for wide-field views of the night sky. This can be swapped for a lens with a longer focal length to provide a higher magnification and therefore capture a smaller patch of the sky. A DSLR is capable of long exposures because its shutter can effectively be locked open. However, any exposure of more than a few seconds will produce an image in which the stars are trailed, appearing as arcs across the sky as the Earth rotates beneath them. For much longer exposures that keep an object centered on the camera sensor, you will need to attach a DSLR to a motorized equatorial mount (see below). The longest exposures demand more specialized, more accurate tracking systems.

LIGHT POLLUTION FILTERS

Light pollution is the bane of the astrophotographer. From most sites, long-exposure photos will show the cast of light emitted by sodium- and mercury-vapor streetlights. Fortunately, these lamps give out light in narrow wavelength ranges; this extraneous light can be removed using dedicated filters designed to block the wavelengths associated with these sources.

ORION REVEALED
The filtered photograph of Orion (right) shows more contrast than the unfiltered image (above right).

PRIME FOCUS
Attaching a camera directly to the telescope tube makes objects larger in the image but requires accurate tracking.

PIGGY-BACKING
A piggy-backed camera will track the motion of the stars as the telescope it rides on moves to follow them.

PORTABLE MOUNT
Small, portable mounts—capable of tracking the sky—are popular for use with DSLR cameras and wide lenses.

Tracking with a DSLR

Long-exposure pictures of celestial objects can only be taken by a camera that can follow the apparent movement of the night sky. This can be achieved with a DSLR by using a portable tracking mount or "piggy-backing" the camera on top of a telescope on a motorized equatorial mount. These method are useful for taking long-exposure shots of large expanses of sky, and can be very effective for capturing nebulae or galaxies that are spread across large areas.

Another technique is to remove the DSLR's lens and attach the camera to a telescope via an adapter, effectively making the telescope into the camera lens. This "prime-focus" technique allows long-exposure (multiminute) close-up shots of galaxies, nebulae, and star clusters, but it requires a mount capable of tracking the night sky with extreme accuracy, and often constant, computer-aided corrections to the motion of the mount. This is a specialized area of astrophotography.

CCD imaging

Charge Coupled Devices (or CCDs) are dedicated astronomy cameras that are used by many astroimagers to take pictures of faint objects such as star clusters, nebulae, and galaxies. A CCD works by converting the light that falls on its chip into an electric signal. This signal is then read by the CCD to form an image of the object being observed. The image is then downloaded via a cable to a computer. CCD chips provide advantages over DSLRs for astrophotography because they are extremely sensitive, particularly to the wavelengths of light given off by celestial objects, and are cooled to reduce thermally generated noise. CCDs are available in monochrome versions—which require use of an additional set of color filters to create a color image—as well as so-called "one-shot-color" models.

CCD IMAGE PROCESSING
The raw data produced by a CCD usually requires a lot of computer processing in order to create a finished picture.

CCD CAMERAS
Some CCDs come with cooling, for long-exposure deep-sky imaging (left), while others are capable of high-frame-rate imaging of the Moon and planets (right).

COOLED CCD

HIGH-FRAME-RATE CCD

COLORED WHIRLPOOL
This stunning image of the Whirlpool
Galaxy, in the constellation Canes
Venatici, was taken using a 360mm
catadioptric telescope and made by
combining separate red, green, and blue
channels in specialized graphics software.

4

PATHFINDERS

Astronomers have divided up the night sky into areas that contain certain groups of stars; each of these 88 groups is known as a constellation. By recognizing the constellations, you can "starhop" from one area to another, and locate spectacular deep-sky objects.

THE CONSTELLATIONS

The constellations help astronomers find their way around the heavens by breaking up the night sky into manageable pieces containing patterns that are easy to spot and remember.

The birth of Western constellations

The idea of "constellations" dates back thousands of years. Paintings in the Lascaux Caves of southern France are believed to be around 16,000 years old and may show one of the earliest depictions of the constellation we now know as Taurus, the Bull. For these early cultures, their local animals, environment, and stars were linked as they were all a part of nature. However, the first real evidence of constellation designs, and indeed the beginnings of astronomy, come from ancient Greek times. In his poem, the *Phaenomena*, Aratus (*c.*315–*c.*245 BCE) describes 47 constellations based on a lost work of an earlier Greek, Eudoxus (*c.*390–*c.*340 BCE). He in turn learned of them from the Egyptians, who probably gained their knowledge from Babylonian, Sumerian, and Phoenician sources. Around 150 CE, the astronomer Ptolemy brought together his own and earlier Greek work in a multi-volume compilation known as the *Almagest*, which included a star catalog and a list of 48 constellations. Well over 1,000 years later this "book" found its way to Italy and was translated into Latin, which is why we have Latin names for the constellations.

Constellations are the product of **human imagination** and the urge to impose order on the **mysteries** of the night sky.

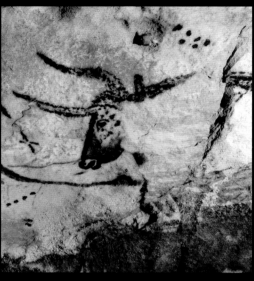

AN EARLY STAR MAP?
The Paleolithic painting of a bull in the caves of Lascaux may represent the constellation Taurus. The set of dots above the shoulder of the bull are thought by some to depict the Pleiades open star cluster.

Chinese constellations

The constellations that western astronomers recognize today came from earlier Greek works and were added to by Western sailors, navigators, and astronomers. However, many other cultures around the world had their own constellations, including the ancient Chinese, who as early as 3000 BCE, divided the sky up into 31 regions of three Enclosures and 28 Mansions. The Enclosures are separated from each other by "walls" formed by asterisms and are located around the north celestial pole. Meanwhile the Mansions are the Chinese equivalent of the western zodiac, except they are based on the monthly movement of the Moon and not the yearly track of the Sun (though the difference between these two paths is not great). There are also four mythical creatures in the Chinese constellations: Azure Dragon of the East (Spring), Vermilion Bird of the South (Summer), White Tiger of the West (Autumn), and the Black Tortoise of the North (Winter); each one is used to represent the indicated direction as well as one of the four seasons in the year.

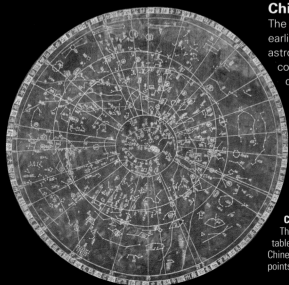

CHINESE CHARTS
This star chart was engraved on a stone tablet in 12th-century China. The system of Chinese astronomy uses different reference points to western astronomy to locate the stars.

Exploration and the constellations

Once Western explorers began to sail to the southern lands, they saw large areas of the night sky that lacked constellations. The Greeks and Romans had viewed the sky from the northern hemisphere, so their 48 ancient constellations did not extend all the way to the southern celestial pole. But it was not long before navigators and astronomers began filling in the blanks by picking out their own constellations. From the late 1500s new constellations appeared, beginning with a set of 12 from Dutch explorers Pieter Keyser and Frederick de Houtman and their mentor Petrus Plancius. The scene was set, and astronomers tried to squeeze in constellations wherever they could—in many cases to honor their sponsors or monarch. For example, in 1679 several stars in Aries were used by Augustin Royer to make the new constellation of Lilium, the Lily, which represented the fleur-de-lis of his benefactor, King Louis XIV of France. The Lily did not survive, and the same happened to many other constellations that were deemed unnecessary. The International Astronomical Union finally fixed the constellations in place during their first meeting in 1922, so the heavens will no longer change.

CONSTELLATION FIGURES

The *Atlas Coelestis* from 1729 shows the classical representations of the original 48 Ptolemaic constellations as well as many others that have since been abandoned.

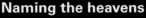

BAYER'S ORION

Johann Bayer's star atlas, *Uranometria* (literally "measuring the sky") contains magnificent engravings of the constellations, such as Orion.

Naming the heavens

The names given to stars and constellations were rationalized by the German astronomer Johann Bayer. In 1603, in homage to earlier Greek astronomers, he assigned the brightest stars of a constellation Greek letters in his star atlas *Uranometria*. So, the first star (usually the brightest, but not always) in any constellation would be alpha, followed by the genitive (or Latin possessive, meaning "belonging to") form of the constellation name. Using Orion as an example, the main star has the Arabic name of Betelgeuse, while the new Bayer designation became Alpha Orionis, or α Orionis. The possessives, as is the case with many languages, need to be learned as they are not simply the constellation name with "is" on the end. For example, the Latin possessive for Gemini is Geminorum and for Pegasus is Pegasi. In addition, all the constellations have a three-letter abbreviation, which make names quicker and easier to write. Orion is simply Ori, so Betelgeuse, again, could also be written as α Ori.

MAPPING THE CONSTELLATIONS

This section of the book maps those constellations with the most recognizable star patterns or which contain the most notable deep-sky objects.

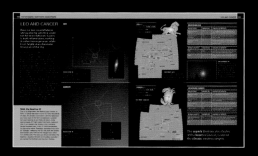

Understanding the constellation maps

The constellation maps in this section are orientated with north at the top and south at the bottom. They are all reproduced to the same scale, to give an accurate reflection of the relative sizes of the constellations. Each map depicts the constellation figure (the pattern of lines joining the brightest stars) and the boundaries of the constellation—as defined by the International Astronomical Union—which are outlined in red. Only those constellations with recognizable star patterns and/or notable deep-sky objects are included in this section. However, a full list of all the 88 officially recognized constellations can be found in the Reference section (see pp.240–241) and on p.77.

CONSTELLATION DESCRIPTIONS
The most prominent and interesting constellations in the night sky are explored on the following pages. Details are given for the brighter and more interesting stars and deep-sky objects in each of them.

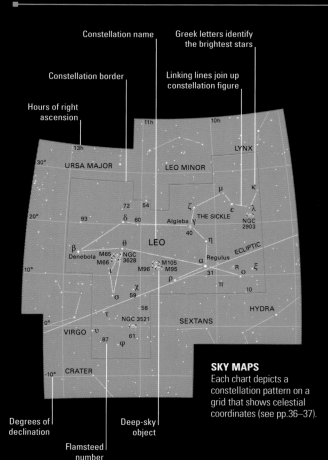

Constellation name

Greek letters identify the brightest stars

Constellation border

Linking lines join up constellation figure

Hours of right ascension

Degrees of declination

Flamsteed number

Deep-sky object

SKY MAPS
Each chart depicts a constellation pattern on a grid that shows celestial coordinates (see pp.36–37).

What the constellation maps show

The maps show the stars in each constellation down to a magnitude of 6.5, just within naked-eye visibility, making them useful when stargazing with the naked eye or a pair of binoculars. Stars with a noticeable color or variable brightness are also indicated on these maps, as are deep-sky objects, such as star clusters, nebulae, and galaxies.

Star magnitudes

-1.5–0 0–0.9 1.0–1.9 2.0–2.9 3.0–3.9 4.0–4.9 5.0–5.9 6.0–6.9 Variable star

Deep-sky objects

Galaxy Open cluster Globular cluster Planetary nebula Diffuse nebula

STARS AND OBJECTS
Major stars are represented by dots that indicate their apparent magnitude.

DEEP-SKY OBJECTS
Icons locate different deep-sky objects; all are labeled with their catalog number.

BAYER LETTERS AND FLAMSTEED NUMBERS

The brightest stars in a constellation are identified by letters of the Greek alphabet, following a catalog invented by Johann Bayer (see p.73). Close pairs or groups of stars may share a Greek letter and are distinguished by superscripts. Other stars are cataloged by a number—for example 15 Orionis—known as the Flamsteed number. They are numbered in increasing order of right ascension (right to left on the sky maps).

α Alpha		ν Nu	
β Beta		ξ Xi	
γ Gamma		ο Omicron	
δ Delta		π Pi	
ε Epsilon		ρ Rho	
ζ Zeta		σ Sigma	
η Eta		τ Tau	
θ Theta		υ Upsilon	
ι Iota		φ Phi	
κ Kappa		χ Chi	
λ Lambda		ψ Psi	
μ Mu		ω Omega	

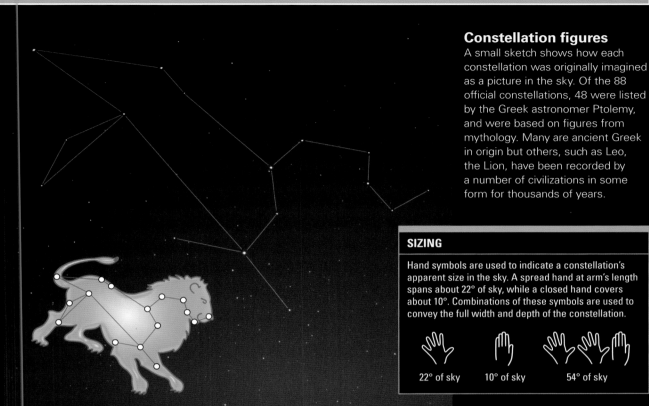

Constellation figures

A small sketch shows how each constellation was originally imagined as a picture in the sky. Of the 88 official constellations, 48 were listed by the Greek astronomer Ptolemy, and were based on figures from mythology. Many are ancient Greek in origin but others, such as Leo, the Lion, have been recorded by a number of civilizations in some form for thousands of years.

SIZING

Hand symbols are used to indicate a constellation's apparent size in the sky. A spread hand at arm's length spans about 22° of sky, while a closed hand covers about 10°. Combinations of these symbols are used to convey the full width and depth of the constellation.

22° of sky	10° of sky	54° of sky

SEEING STARS

Beside every photograph shown in this chapter is a small icon, which indicates what equipment (if any) is needed to see the pictured object. For example, objects that can be seen with the naked eye alone will be tagged with the 👁 icon, while those that require a telescope to be visible are marked with the 🏹 icon.

Visibility icons

👁 Naked eye

🔭 Binoculars

🏹 Telescope

⚲ Professional observatory

M65 🏹
This spiral galaxy in Leo requires an amateur telescope to be seen and is accordingly labeled with the 🏹 icon.

Visible constellations

Which constellations you are able to observe depends on the latitude of your position on Earth (see p.33). Far-southern constellations cannot be seen from far-northern latitudes, for example, because they will never rise above the horizon. Each constellation's information table gives the range of latitude from which it is fully visible. The map below can be used to find the geographical area that corresponds to these latitudes. The southern-hemisphere constellation Sagittarius, for example, is fully visible from 44°N to 90°S. A stargazer observing from a latitude above 44°N will therefore be unable to make out the whole of this constellation. However, some constellations, such as Orion, span the celestial equator, and can be seen from virtually the entire inhabited world.

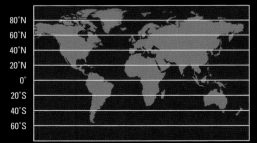

FULLY VISIBLE
Each constellation's entry gives the range of latitudes from which it can be fully seen. This simple map (left) shows which latitudes correspond to different parts of the world. If a constellation is close to an observer's north or south limit, it will never rise far above the horizon, where it is difficult to see.

SKY MAPS

The charts on pp.76–81 form a complete map of the night sky, setting the positions of the constellations in the context of the whole celestial sphere.

About the charts

On the following pages, the celestial sphere has been divided into six sections—two covering the northern and southern polar regions, and the remaining four covering the equatorial areas. When gazing into the night sky, we seem to be viewing part of a great dome that surrounds us, but to view the sky in a book, it must obviously be flattened. This means that some of the constellations appear slightly stretched on paper when compared to their actual appearance, but this is only really noticeable around the edges of the north and south polar sky charts.

The Pathfinders chapter continues after these charts on pp.82–171 with a more detailed look at selected individual constellations, offering objects for the practical astronomer to observe, such as nebulae, star clusters, galaxies, and stars of note, as well as suggested routes, or "starhops," around the night sky.

SIZE MATTERS
Constellations vary greatly in size and prominence; the largest, Hydra, is as wide as 172 full Moons side by side; despite being the smallest, Crux (above) is one of the most distinctive in shape.

THE NORTH POLAR SKY

Sitting less than 1° from the north celestial pole is the celebrated star Polaris, which also goes by the popular names of the North or Pole Star. This is not a particularly bright star, but it is important in allowing navigators on land or sea to find north, and hence their direction of travel. As the Earth's axis points to a spot very close to Polaris, the star appears to stay in virtually the same place as the sky rotates around it. Also, the angular height at which Polaris lies above the horizon is equal to the latitude of an observer in the northern hemisphere.

CELESTIAL SPHERE

Star magnitudes

-1	0	1	2	3	4	5

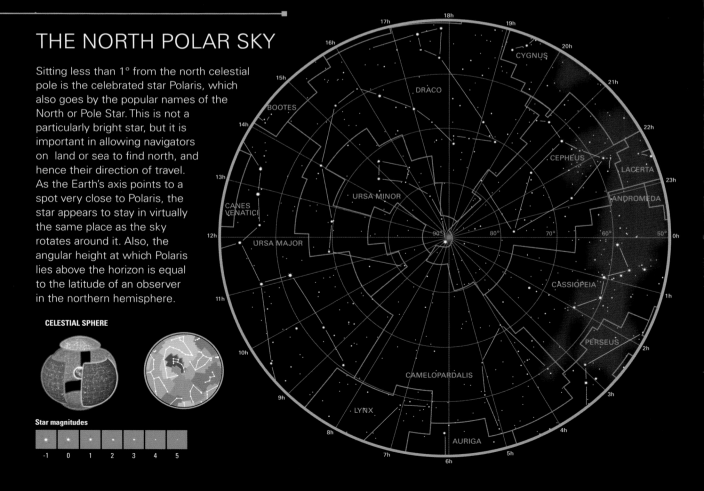

ALPHABETICAL INDEX OF THE 88 CONSTELLATIONS

THE SOUTH POLAR SKY

It is evident from a glance at the south polar chart that while the northern hemisphere has Polaris to guide travelers, there is no equivalent "South Star" in the southern hemisphere. Here, observers must use a fairly complicated method to find the south celestial pole—using stars from the constellations of Crux and Centaurus (see p.140)—as it lies in an otherwise faint and barren part of the night sky. However, the constellations nearby contain some wonderful stars, and are dominated by the incredible band of the Milky Way, with its associated nebulae and star clusters.

CELESTIAL SPHERE

Star magnitudes

-1 0 1 2 3 4 5

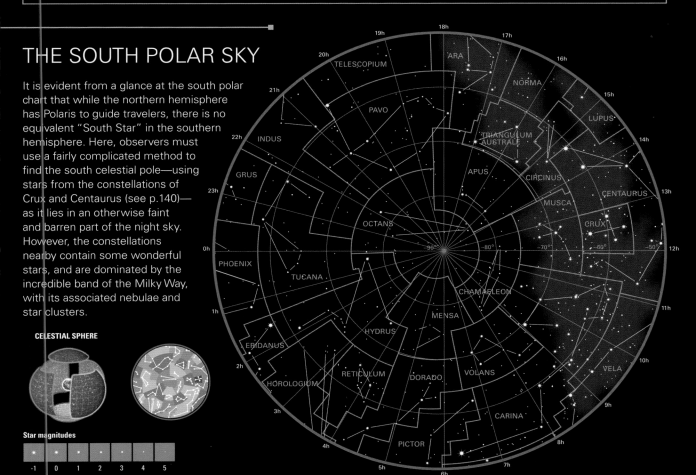

EQUATORIAL SKY CHART 1

This section of the night sky is best placed for worldwide observation during the evenings of September, October, and November. The region is home to the vernal equinox (see p.121), and mostly consists of empty regions of space and faint stars. However, there are still some fine objects to observe. In the constellation of Andromeda you can observe the astounding M31, the Andromeda Galaxy, the largest neighboring galaxy to our own. Just north of the celestial equator is the Great Square of Pegasus, an asterism that acts as a useful signpost around the region. In Cetus lies the marvellous red variable star, Mira. Finally, in Piscis Austrinus to the south is Fomalhaut, the only bright star in this region.

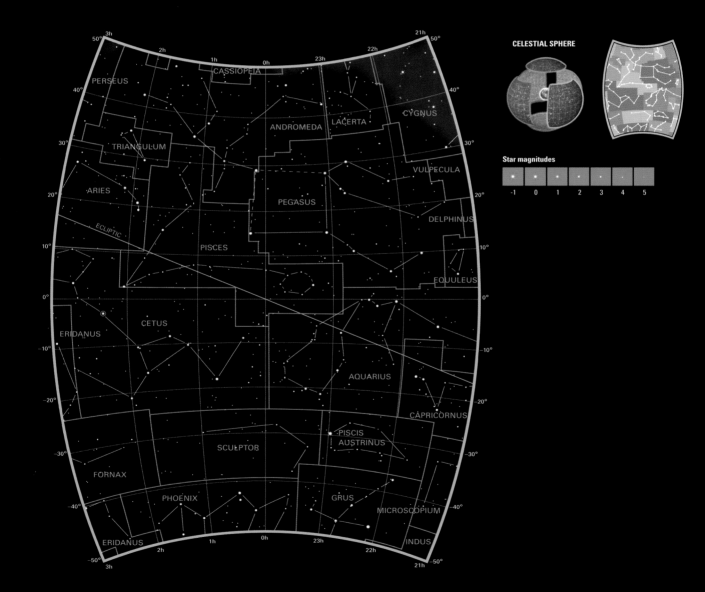

CELESTIAL SPHERE

Star magnitudes

-1	0	1	2	3	4	5

EQUATORIAL SKY CHART 2

This part of the sky is best placed to be observed worldwide during the evenings of June, July, August, and September. Unlike the previous chart, the area here is filled with notable star patterns, and for those living in darker locations the rich starfields of the Milky Way band are here seen at their best. The northern area is dominated by the famous Summer Triangle asterism, comprising Deneb, in Cygnus; Vega, in Lyra; and Altair, in Aquila. To the south is the wonderful curving arrangement of stars that forms the constellation of Scorpius, and nearby Sagittarius contains the recognizable Teapot asterism. To the north of the chart, outside the Milky Way band, is the constellation of Hercules, in which the Great Globular Cluster, M13, can be found.

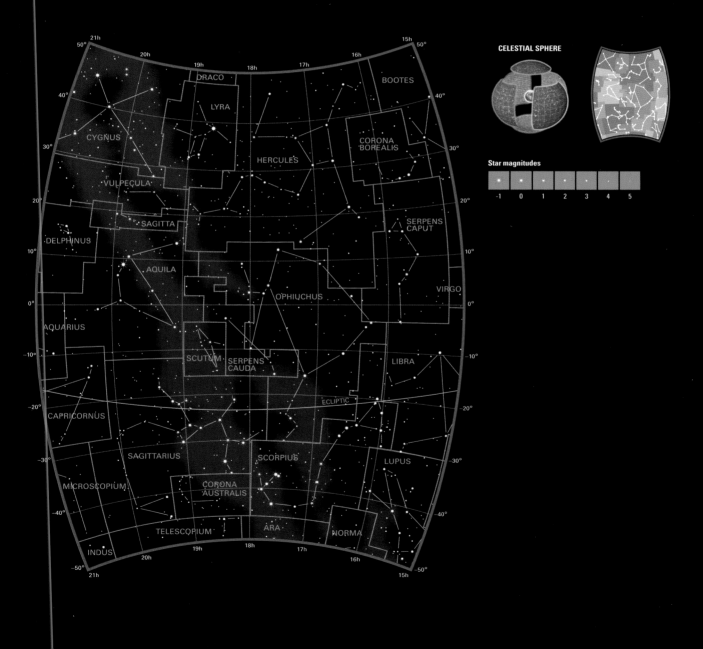

CELESTIAL SPHERE

Star magnitudes

-1	0	1	2	3	4	5

EQUATORIAL SKY CHART 3

This section of the night sky is best observed worldwide during evenings in March, April, and May. The region contains the third- and fifth-brightest stars of the entire celestial sphere: the reddish Arcturus in Boötes—which becomes visible in the northern-hemisphere spring—and the bluish Spica in Virgo respectively. Virgo is also the second-biggest constellation, and just to its south is the largest, Hydra, which weaves its way across—and indeed, beyond—the entire region. Just above the center and to the right are the stars that make up Leo, the Lion. This is one of the few constellations with an easily recognizable pattern, its head forming a distinctive backward "question mark" of six stars.

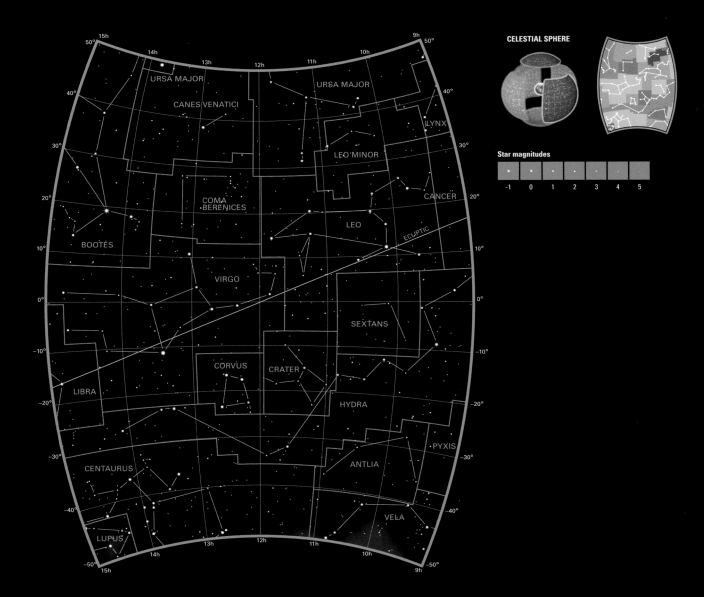

CELESTIAL SPHERE

Star magnitudes

-1   0   1   2   3   4   5

EQUATORIAL SKY CHART 4

This section of the night sky is best placed for observation worldwide during the evenings of December, January, and February. It is one of the most stunning areas of the heavens, simply because it contains more bright stars than any other. The region is home to the distinctive constellation of Orion, with an easily recognizable line of three almost-equally bright stars across its center, known as Orion's Belt. Adjacent to Orion is Taurus, which contains the bright star Aldebaran, as well as one of the celestial sphere's finest star clusters, M45—more commonly known as the Pleiades or the Seven Sisters. Below center is the constellation of Canis Major, which is home to Sirius, the Dog Star—the brightest star in the night sky.

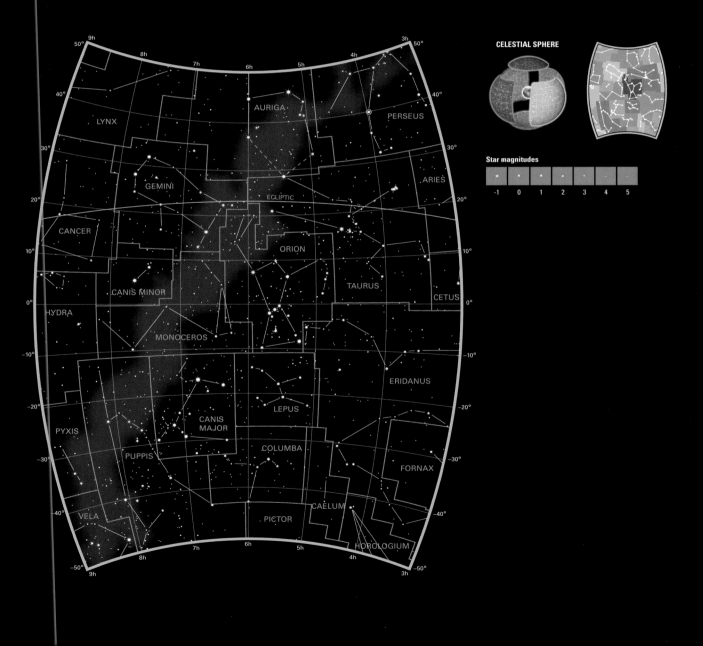

CELESTIAL SPHERE

Star magnitudes

| -1 | 0 | 1 | 2 | 3 | 4 | 5 |

URSA MAJOR

Ursa Major, the Great Bear, contains a group of seven stars known as the Big Dipper, or Plow, one of the best-known and most easily identified star patterns in the night sky.

Ursa Major is the third largest constellation of them all, only beaten by Virgo and, largest of all, Hydra, the Water Snake. However, Ursa Major trumps its larger rivals for interest with a combination of some easily viewed deep-sky objects—such as the galaxy M81 and a fine unaided-eye double star, Mizar—and its most important feature, the bright asterism called by many names, including the Wagon, the Saucepan, the Big Dipper, and the Plow (or Plough). All of the names readily describe how this group looks in the sky, and in fact it is so identifiable that it has become synonymous with the whole constellation. This is incorrect, as the entire pattern of Ursa Major itself takes on the figure of a bear.

Stars within the **Great Bear** point the way toward the **north celestial pole**.

MIZAR AND ALCOR 👁

THE BIG DIPPER

M81, BODE'S GALAXY ♖

Mizar and Alcor 👁
The two stars Zeta (ζ) Ursae Majoris and 80 Ursae Majoris, Mizar and Alcor, make up the famous double star, visible to the naked eye, that sits at the central bend of the Big Dipper's handle. This is not a true binary, as the stars are not linked gravitationally. However, through a small telescope, Mizar is revealed to have a closer companion 15″ away, and these two are a true binary pair, orbiting one another every 5,000 years or so. This is regarded as one of the finest multiple star systems in the night sky. The view is enhanced by the chance alignment of a reddish, magnitude 8.8 star named Sidus Ludovicianum in 1772 by astronomer Johann Liebknecht, who believed he had found a new planet.

WIDE VIEW 👁
Ursa Major in its entirety covers a large area of the sky. Some of its stars are faint, but its overall pattern can be identified from a darker location.

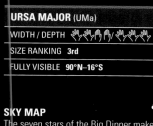

URSA MAJOR (UMa)		
WIDTH / DEPTH		
SIZE RANKING	**3rd**	
FULLY VISIBLE	90°N–16°S	

SKY MAP

The seven stars of the Big Dipper make up the body and tail of the Bear. Two of them, the Pointers, can be used to locate the Pole Star (see Starhopping from The Big Dipper, overleaf).

OBSERVING URSA MAJOR

MAJOR STARS	MAGNITUDE	FEATURES
Alpha (α) Ursae Majoris *Dubhe*	1.8	A red giant that sits about 125 light-years away from us. This and Beta (β) Ursa Majoris, Merak, are the Pointers (see Skymap, above) in the Big Dipper.
Epsilon (ε) Ursae Majoris *Alioth*	1.8	The brightest star in the constellation.
Zeta (ζ) Ursae Majoris *Mizar*	2.3	Forms an optical double with the mag. 4.0 star Alcor, 12' away.

NOTABLE OBJECTS	MAGNITUDE	FEATURES
M81, Bode's Galaxy	6.9	A spiral galaxy that is the main member of a group of 34 galaxies in this constellation.
M82, the Cigar Galaxy	8.4	A spiral galaxy of prolific starburst activity due to gravitational interaction with M81.
M97, the Owl Nebula	9.8	A planetary nebula; a very faint object without a larger telescope, in which its owl-like eyes can be seen.
M101, the Pinwheel Galaxy	8.3	A spiral galaxy; only larger telescopes will begin to show its pinwheel formation.

M101

Forming a triangle with Alcor and Alkaid, the last two stars of the Big Dipper's handle, the Pinwheel Galaxy appears head-on to observers on Earth, allowing its spiral arms to be seen through a powerful telescope. This galaxy is estimated to be twice the size of the Milky Way.

M81 and M82

Also known as Bode's Galaxy, M81 is seen in binoculars as a fuzzy patch located just above the Bear's shoulders. This is the closest galaxy to us outside our Local Group, at a distance of just 12 million light-years, which makes it a very bright and large object. For this reason, as well as the fact that there is a second galaxy, M82, sitting just to its north, M81 is a particularly good target for amateur galaxy watchers. M82 is known as the Cigar Galaxy. It is fainter than M81 but still very much visible. These two really are close in space, too; detailed images of M82 show that it has been scarred in a gravitational battle with M81.

GALAXIES IN CONFLICT
Seen side-on from Earth, M82 (above left) has a cigar shape. Up close (right), it can be seen to have been deformed by the gravitational power of M81 (above right), spewing out red filaments that may indicate violent activity within the galaxy.

STARHOPPING FROM THE BIG DIPPER

The Big Dipper is a good signpost, as it sits close to the sky's north pole. For many stargazers in the northern hemisphere, it never dips below the horizon.

The Big Dipper is an asterism of seven stars that forms part of the larger constellation of Ursa Major (the Great Bear). It is helpful to think of the asterism as being shaped like a saucepan, with a handle attached to a bowl.

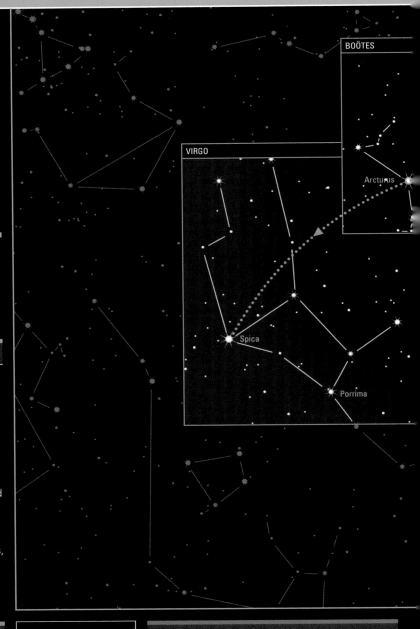

1 TO URSA MINOR AND THE POLE

One of the most important starhops in the northern hemisphere takes you from the Big Dipper to the North Star, Polaris, which marks the position of the celestial north pole. Draw a line between the last two stars in the "bowl," from Merak to Dubhe, the main star of Ursa Major. Extend the line between these stars—known as "the Pointers"— until you reach Polaris, in Ursa Minor.

URSA MINOR

2 TO BOÖTES THEN VIRGO

The Big Dipper's handle is not straight, but bends down from Mizar, the second star from the handle's end. Visualize a curved line connecting the last three stars of the handle and extend this curve to reach Arcturus, the golden leading star of Boötes, and the fourth-brightest star in the night sky. Continue the original arc for the same distance again to reach another, slightly blue,

3 TO LEO

Starting at Megrez (magnitude 3.3), the faintest star of the Big Dipper's seven, draw an imaginary line to Phecda (also known as Phad) at the base of the Big Dipper's "bowl." Continue this line out of Ursa Major and you will pass through the small and relatively insignificant constellation of Leo Minor before reaching the highly distinctive crouching lion pattern of Leo and its bright main

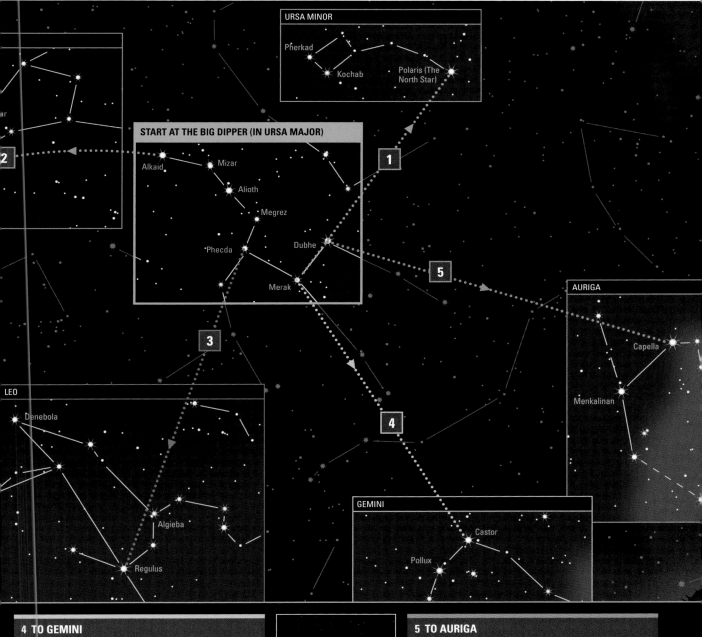

URSA MINOR

Pherkad

Kochab

Polaris (The North Star)

START AT THE BIG DIPPER (IN URSA MAJOR)

Alkaid · Mizar · Alioth · Megrez · Phecda · Dubhe · Merak

1

2

3

5

4

AURIGA

Capella

Menkalinan

LEO

Denebola

Algieba

Regulus

GEMINI

Castor

Pollux

4 TO GEMINI

Mizar, the point at which the handle of the Big Dipper bends, is the start of a journey to Gemini. Extend an imaginary line from Mizar through Alioth and Megrez on the handle of the Big Dipper. This line will pass through Merak on the bowl of the Big Dipper; continuing the line will lead you to two bright stars, Castor (magnitude 1.9) and the warmer-colored Pollux (magnitude 1.2), which are

5 TO AURIGA

The two stars that make the top of the bowl of the Big Dipper are the faint Megrez and the bright Dubhe. Extending a line between the two in the opposite direction to the handle will point you to the 11th-brightest star in the sky, Capella, the leading star in Auriga, the Charioteer. The sky around this starhop is quite empty, filled only with the fainter stars that make the head of the Great

URSA MINOR AND DRACO

These two neighboring constellations have one thing in common: their leading star has at one time been designated the Pole Star.

Neither Ursa Minor, the Little Bear, nor Draco, the Dragon, can be thought of as conspicuous constellations. None of their stars is brighter than 2nd magnitude, which makes viewing them in light-polluted areas somewhat difficult. However, in darker skies, they can be identified readily, as both constellations do have a clear pattern to their stars. In the case of Ursa Minor, it appears similar to a smaller version of the Big Dipper in Ursa Major, while Draco's form weaves gently across the sky, from Ursa Major over Boötes to Hercules and Lyra. Draco is a large constellation that winds nearly halfway around the north celestial pole. Despite this, however, it boasts few sky objects of interest, although it does have an annual meteor shower, the Draconids, which peaks on October 9. Nu (ν) Draconis, in the "head" of the Dragon, is a great double star for binoculars. Draco's other highlight is NGC 6543, the Cat's Eye Nebula, a planetary nebula seen as a faintly greenish "egg" in a small telescope. Ursa Minor contains several double stars, but its main attraction is the star Alpha (α) Ursae Minoris, Polaris, popularly known as the North or Pole Star.

The Pole Star 👁

Many believe that if a brilliant star is spotted, it must be the Pole Star. Unfortunately, this is a myth, as Polaris's magnitude is but a mere 2.0, placing it at the 49th-brightest of the sky's stars. This is not an insignificant star, however, having a diameter 30 times that of the Sun's. Interestingly, Draco's leading star, Thuban, was the Pole Star several thousand years ago. The change is due to the 25,800-year wobble of the Earth's axis, known as precession, which constantly changes where our North Pole is "pointing" in space (see also p.60). For this reason, the term "Pole Star" is only a temporary designation for whichever star happens at that time to lie at or close to the celestial north pole.

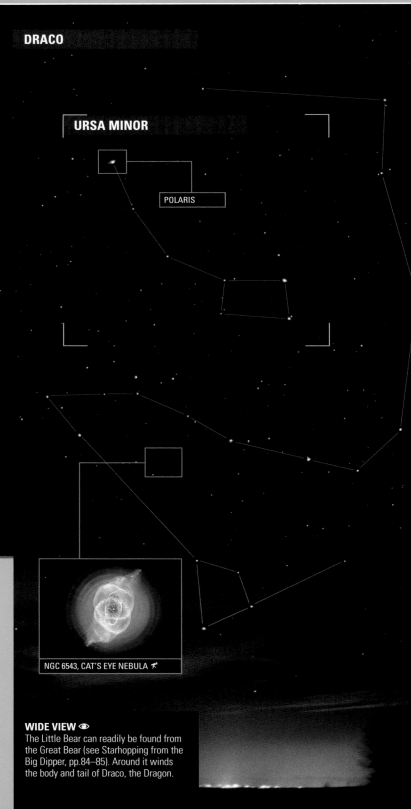

DRACO

URSA MINOR

POLARIS

NGC 6543, CAT'S EYE NEBULA ✦

WIDE VIEW 👁
The Little Bear can readily be found from the Great Bear (see Starhopping from the Big Dipper, pp.84–85). Around it winds the body and tail of Draco, the Dragon.

URSA MINOR (UMi)

WIDTH / DEPTH	
SIZE RANKING	56th
FULLY VISIBLE	90°N–0°S

OBSERVING URSA MINOR

MAJOR STARS	MAGNITUDE	FEATURES OF INTEREST
Alpha (α) Ursae Minoris *Polaris*	2.0	A pale-yellow-colored star at a distance of around 430 light-years from us.
Gamma (γ) Ursae Minoris *Pherkad*	3.0	A relatively bright optical double—with a mag. 5.0 companion, 11 Ursae Minoris—that is visible to the naked eye.

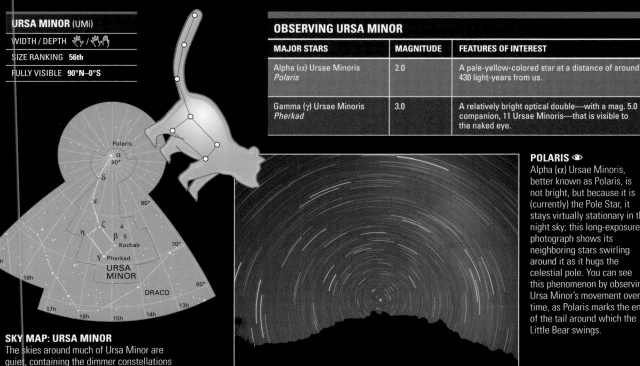

SKY MAP: URSA MINOR
The skies around much of Ursa Minor are quiet, containing the dimmer constellations Camelopardalis, Draco, and north Cepheus.

POLARIS 👁
Alpha (α) Ursae Minoris, better known as Polaris, is not bright, but because it is (currently) the Pole Star, it stays virtually stationary in the night sky; this long-exposure photograph shows its neighboring stars swirling around it as it hugs the celestial pole. You can see this phenomenon by observing Ursa Minor's movement over time, as Polaris marks the end of the tail around which the Little Bear swings.

DRACO (Dra)

WIDTH / DEPTH	
SIZE RANKING	8th
FULLY VISIBLE	90°N–4°S

SKY MAP: DRACO
Draco is the eighth-largest constellation in the skies, its long tail separating the Great Bear and Little Bear.

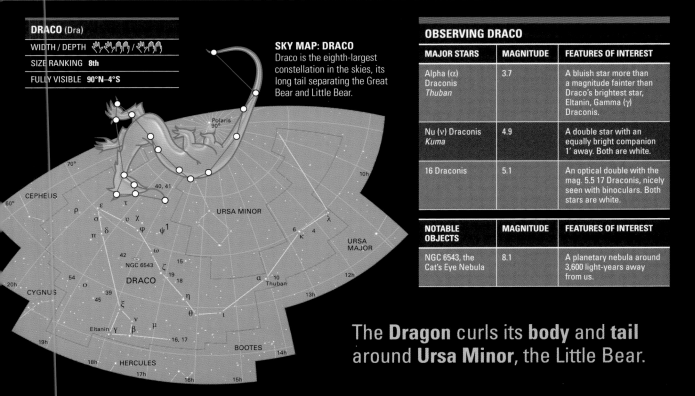

OBSERVING DRACO

MAJOR STARS	MAGNITUDE	FEATURES OF INTEREST
Alpha (α) Draconis *Thuban*	3.7	A bluish star more than a magnitude fainter than Draco's brightest star, Eltanin, Gamma (γ) Draconis.
Nu (ν) Draconis *Kuma*	4.9	A double star with an equally bright companion 1' away. Both are white.
16 Draconis	5.1	An optical double with the mag. 5.5 17 Draconis, nicely seen with binoculars. Both stars are white.

NOTABLE OBJECTS	MAGNITUDE	FEATURES OF INTEREST
NGC 6543, the Cat's Eye Nebula	8.1	A planetary nebula around 3,600 light-years away from us.

The **Dragon** curls its **body** and **tail** around **Ursa Minor**, the Little Bear.

CASSIOPEIA

Cassiopeia, the Queen, has everything a constellation could ask for: a fine, distinctive shape and many objects to look for, as she sits right in the center of the band of the Milky Way.

There are just a handful of constellations, or parts thereof, which can claim to be instantly recognizable: Orion, the Big Dipper, and Crux, for example, and also Cassiopeia. Her five main stars form a wonderful W-shape (or M-shape, depending on the view) that from darker locations stands out against the Milky Way. Cassiopeia sits on virtually the opposite side of the north celestial pole to the Big Dipper of Ursa Major. This means, for the majority of northern-hemisphere viewers, that when Cassiopeia is high, the Big Dipper is near the horizon, and vice versa. In addition, due to its high northern declination, the constellation is circumpolar from many places, which means that it is always visible somewhere in the sky and never sets below the horizon.

In Greek legend, **Cassiopeia** is **condemned** to circle the **celestial pole** forever.

NGC 457, the Skiing Cluster

Anything out there with a pattern of stars is of real benefit to the amateur stargazer, because once spotted, it is easy to identify. In the case of the open star cluster NGC 457, most of the brighter members form the entertaining shape of a skier with two bright eyes, holding up ski poles. Others have likened the shape to an owl or even to the movie character E.T. waving his arms. The brightest star in this group, Phi (φ) Cassiopeiae (the right "eye"), is not actually a true cluster member but a star that is 6,700 light-years closer to us. This confirms that everything that is seen in the night sky is so far away that it all just appears to be at the same "fixed" distance, while the gaps between objects can be vast.

GAMMA (γ) CASSIOPEIAE 👁

NGC 457, SKIING CLUSTER 🔭

WIDE VIEW 👁
The distinctive W-shape of Cassiopeia's main stars makes it one of the easiest constellations to locate in the night sky of the northern hemisphere.

CASSIOPEIA (Cas)	
WIDTH / DEPTH	🖐🖐 / 🖐🖐
SIZE RANKING	25th
FULLY VISIBLE	90°N–12°S

M103 👀

The M103 open star cluster sits just below the Queen's right (crossed) knee. Its most recognizable feature is a chain of three stars that closely resembles a miniature version of Orion's belt (see pp.134–135).

OBSERVING CASSIOPEIA

MAJOR STARS	MAGNITUDE	FEATURES
Alpha (α) Cassiopeiae *Schedar*	2.2	An orange giant that is the brightest star in the constellation, except when the variable Gamma (γ) Cassiopeiae (below) is at its brightest.
Gamma (γ) Cassiopeiae *Tsih*	1.6–3.0	An unstable, eruptive variable that spins at a colossal rate, causing gas to be thrown off periodically.

NOTABLE OBJECTS	MAGNITUDE	FEATURES
M52	7.3	Binoculars will show this open star cluster as a misty patch with a brighter orange star off to one side.
M103	7.4	A lovely open star cluster crossed by a chain of three colored stars.
NGC 457, the Skiing Cluster	6.4	A 9,000-light-year distant open star cluster; a good target for binoculars.
NGC 663	7.1	A fine open star cluster for binoculars, but even better seen in a small telescope.

SKY MAP

The south of Cassiopeia contains the bright stars and Milky Way; the north is a much quieter area.

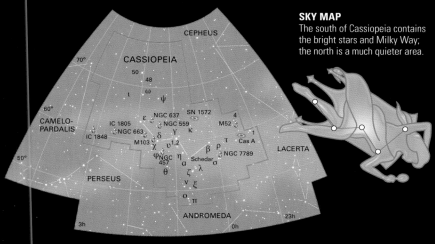

NGC 663 👀

Many of the stars in this cluster are, like Gamma (γ) Cassiopeiae, known to be spinning incredibly fast, throwing off (principally hydrogen) gas.

M52 and the Milky Way 👀

M52 is a family of over 180 stars that sits 5,000 light-years away. Easily picked out in binoculars, it is a great sight for a small telescope, in which the entire scene will be filled with other stars of the Milky Way. Like M52, virtually all open star clusters are found within the Milky Way band, simply because the dust and gas here creates the majority of the star clusters we can see. Our galaxy is a huge, wheel-shaped spiral of stars and nebulae. When we look out, these stars and nebulae are at their most dense looking "across" the wheel, forming a band that seems to encircle Earth. Star clusters form within nebulae, hence most appear in the Milky Way.

SIBLING STARS

Like all open clusters, M52 is made up of "sibling" stars of similar age, formed from the same nebulous clouds of interstellar dust and gases.

CEPHEUS AND CAMELOPARDALIS

Cepheus, the King, and Cameloparalis, the Giraffe, are constellations of the far north. They meet at, and are virtually on opposite sides of, the sky's north celestial pole.

Through the southern portion of Cepheus flows the northernmost point of the starry band of our galaxy known as the Milky Way. If that is visible then generally so are objects worth viewing. Cepheus contains a very special variable star, Delta (δ) Cephei, that has contributed hugely to the study of variables (see below), but from a stargazer's perspective, the exceptional object here is IC 1396, an enormous nebula found around and to the south of the star Mu (μ) Cephei, the Garnet Star. This may just be viewed with the unaided eye, but a large telescope is needed to bring out any of the detail in this large and complex star-forming cloud. It sits south of the base of a slightly stretched "house" pattern formed by the main stars of the constellation. The roof of this house points toward the constellation of Camelopardalis, but do not expect to see anything remarkable here, as no star in Camelopardalis is brighter than mag. 4.0. There is one main object of interest here: a chain of stars called Kemble's Cascade.

Delta Cephei 👁

This is one of the most important variable stars, for it was used to establish a law that helped discover the distance of remote objects. It was found that the period over which the brightness of this pulsating star varied was linked to its intrinsic brightness, or luminosity. The same law applied to all other stars of its type, so they are named Cepheid variables. According to another equation, stars look fainter the further away they are, so if in some galaxy a star is seen like Delta Cephei, its real luminosity can be calculated and the distance/brightness equation can then be used to work out how far away it is. This process was used by Edwin Hubble to prove that the Andromeda Galaxy lay outside the Milky Way.

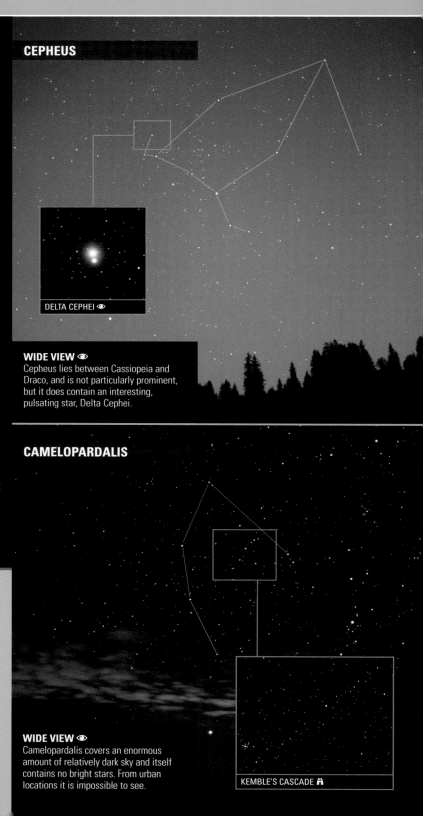

CEPHEUS

DELTA CEPHEI 👁

WIDE VIEW 👁
Cepheus lies between Cassiopeia and Draco, and is not particularly prominent, but it does contain an interesting, pulsating star, Delta Cephei.

CAMELOPARDALIS

WIDE VIEW 👁
Camelopardalis covers an enormous amount of relatively dark sky and itself contains no bright stars. From urban locations it is impossible to see.

KEMBLE'S CASCADE 🔭

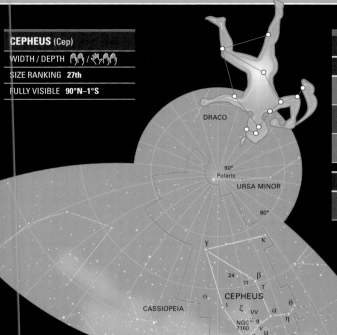

CEPHEUS (Cep)

WIDTH / DEPTH	
SIZE RANKING	**27th**
FULLY VISIBLE	**90°N–1°S**

OBSERVING CEPHEUS

MAJOR STARS	MAGNITUDE	FEATURES OF INTEREST
Alpha (α) Cephei *Alderamin*	2.5	A white star currently evo a red giant.
Delta (δ) Cephei *Al Radif*	3.5–4.4	A famous variable star wit 5 days 9 hours.
Mu (μ) Cephei *Garnet Star*	3.4–5.1	A red supergiant; one of th known.

NOTABLE OBJECTS	MAGNITUDE	FEATURES OF INTEREST
IC 1396	6.0	An emission nebula about light-years away from us.

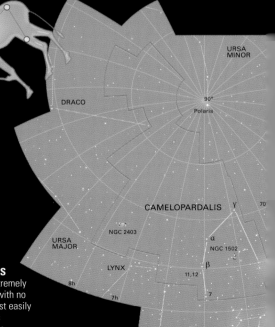

IC 1396
The blandly nar
cluster of stars
the most extens
nebula visible.
3 degrees, whic
the diameter of
Looking over th
the massive re
Cephei (far righ
diameter approx
times wider tha

SKY MAP: CEPHEUS

Cepheus is not immediately conspicuous, but if Cassiopeia is visible, look "above" its W-shape.

CAMELOPARDALIS (Cam)

WIDTH / DEPTH	
SIZE RANKING	**18th**
FULLY VISIBLE	**90°N–3°S**

OBSERVING CAMELOPARDALIS

MAJOR STARS	MAGNITUDE	FEATURES OF INTEREST
Alpha (α) Camelopardalis	4.3	A blue "supergiant" star with a diameter 29 times that of the Sun.

NOTABLE OBJECTS	MAGNITUDE	FEATURES OF INTEREST
Kemble's Cascade	7.0	This asterism is a beautiful binocular target consisting of a chain of around 20 totally unrelated stars.

The gentle **Giraffe** Camelopardalis has the **Dragon, Great Bear,** and **Lynx** as its neighbors

SKY MAP: CAMELOPARDALIS

The Giraffe is an extremely faint constellation, with no bright stars. It is most easily found by locating its neighbor Ursa Major

CANES VENATICI

Canes Venatici, the Hunting Dogs, is a constellation without a memorable pattern of stars to identify it, but it does house the great Whirlpool Galaxy, M51, as well as an interesting variable star called La Superba.

Canes Venatici lies between Boötes and Ursa Major in the northern sky. It was created by the 17th-century astronomer Johannes Hevelius, who utilized stars that had previously been part of Böotes. The constellation is led by the star once grandly known as Cor Caroli Regis Martyris, meaning "the Heart of King Charles the Martyr," after Charles I of England. Its full scientific name is Alpha (α) Canum Venaticorum, an equally splendid designation given that it is verging on 3rd magnitude, in keeping with the quiet visual nature of the rest of the constellation. However, the Hunting Dogs can be found easily, for they sit a short distance south of the Big Dipper's handle. Just within Canes Venatici's southern boundary with Coma Berenices is M3, one of the northern hemisphere's best globular clusters and an object visible with the unaided eye from really dark locations. Another notable deep-sky object is found near the border with neighboring Ursa Major: the famous Whirlpool Galaxy, M51. This can be seen with binoculars or a small telescope as an oval, misty patch of sky.

COR CAROLI ⚹

LA SUPERBA

M51, THE WHIRLPOOL GALAXY ⚹

La Superba 👥
The variable star Y Canum Venaticorum was named "the Superb" by the Italian astronomer Father Angelo Secchi, who, while observing its changes in brightness was impressed by its deep red color. The star sits at a distance of 710 light-years away from the Solar System, and, as with all red variable stars, La Superba is an example of what happens as a star nears the end of its main sequence lifetime (see p.17) and runs out of fuel. Calculations show that this star has grown to a diameter equivalent to four times the distance from the Earth to the Sun, and its brightness fluctuates every 160 days. It is also one of the coldest known stars, with a surface temperature of around just 3,600°F (2,000°C).

WIDE VIEW 👁
Cor Caroli and the fainter Beta (β) Canum Venaticorum, Chara, are the only main designated stars in the constellation of Canes Venatici.

CANES VENATICI (CVn)

WIDTH / DEPTH	/
SIZE RANKING	**38th**
FULLY VISIBLE	**90°N–37°S**

SKY MAP
Canes Venatici occupies the empty triangular patch of sky between Ursa Major, Boötes, and Leo. Until the 17th century it was part of Boötes.

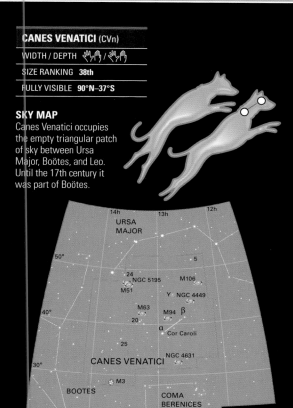

OBSERVING CANES VENATICI

MAJOR STARS	MAGNITUDE	FEATURES
Alpha (α) Canum Venaticorum *Cor Caroli*	2.9	A white binary star, 2.5 times the diameter of the Sun, with a mag. 5.6 companion.
Beta (β) Canum Venaticorum *Chara*	4.3	A yellow star very similar to the Sun in terms of mass, age, and evolutionary status.
Y Canum Venaticorum *La Superba*	4.8–6.4	A variable star which changes brightness over 160 days.

NOTABLE OBJECTS	MAGNITUDE	FEATURES
M3	6.2	A globular cluster found approximately halfway between the stars Arcturus and Cor Caroli.
M51, the Whirlpool Galaxy	8.4	A spiral galaxy that cannot be observed in light-polluted skies, so the darker your site the better.

M3
About 34,000 light-years from Earth, this is one of the biggest globular clusters in the northern sky, with around half-a-million stars. It is a fine sight in a small telescope, where it appears about half the size of the full Moon, but is also just visible to the naked eye. A larger telescope with an aperture of 4 inches (100mm) is needed to see its individual stars.

The two hunting dogs Canes Venatici strain on the **leash** of the **Herdsman**, Boötes.

M51, the Whirlpool Galaxy

This was once known as the "Whirlpool Nebula" before telescopes became larger and better-constructed in the mid-19th century. One of the finest of the new generation was the 72in (1.8m) Leviathan built by William Parsons in 1845. That year he became the first person to witness, and draw, the star-filled and spiral structure of M51, a discovery that was repeated with many other objects that were previously thought to be mere clouds of dust and gas. The Whirlpool Galaxy consists of two interacting galaxies: a spiral galaxy that is observed almost head-on, and a smaller, irregular galaxy. The Whirlpool is similar in size to the Milky Way.

LUMINOUS WHIRLPOOL
The Whirlpool Galaxy needs a larger telescope to reveal its spiral-arm structure. It appears as a misty oval in a small amateur telescope.

BOOTES AND COMA BERENICES

Boötes, the Herdsman, is the brighter of these neighboring constellations, and the easier to recognize. However, Coma Berenices is also worth finding, for its notable globular cluster and several open star clusters.

Boötes is a prominent constellation that extends from Draco to Virgo, presided over by its powerful main star Alpha (α) Boötis, or Arcturus. This is a red giant that is also the brightest star in the northern hemisphere and the fourth-brightest overall. One of the best ways to locate Arcturus is to extend an imaginary curve out from that created by the handle of the Big Dipper, the famous seven-star pattern within Ursa Major. It will be obvious when this star-hop has arrived at Arcturus, because of its brightness and also its golden coloring. Extending from Arcturus are the five main stars of Boötes; they are all fairly bright and form an easily recognizable kite shape in the night. However, apart from Arcturus, this is a rather quiet area of the sky that includes the faint constellation of Coma Berenices, the Hair of Berenice. The brightest star in this constellation is Beta (β) Comae Berenices, which at magnitude 4.3, compares rather poorly with the blazing -0.04 of Arcturus.

M53 ⚲

Of the Milky Way's 158 globular star cluster companions, the M53 cluster is one of the most distant to be found, sitting around 60,000 light-years from the center of the Milky Way and 58,000 light-years from the Solar System. The M53 globular cluster can be found close to the the star Alpha (α) Coma Berenices, or Diadem (symbolizing Berenice's "crown"). It appears in a small telescope as a slightly oval, nebulous object with a large, bright center. Globulars are vast, old systems of stars, and it seems that they are some of the first stars to be created in a new galaxy. Astronomers do not know how or why they are made. It is one of the many unsolved mysteries of the Universe.

BOOTES

ARCTURUS

ALKALUROPS ⚲

WIDE VIEW 👁
The major stars of the constellation of Boötes form a kite shape, with the leading star Alpha Boötis, Arcturus, at its southernmost tip.

COMA BERENICES

M53 ⚲

WIDE VIEW 👁
The three stars that form the constellation can be tricky to find. However, if you can locate the star Arcturus, in Boötes, Coma Berenices is just a short distance away.

SKY MAP: BOOTES

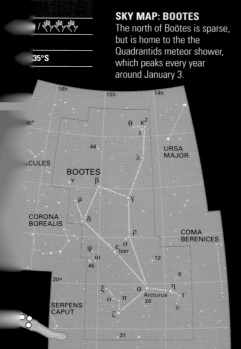

35°S

The north of Boötes is sparse, but is home to the the Quadrantids meteor shower, which peaks every year around January 3.

OBSERVING BOOTES

MAJOR STARS	MAGNITUDE	FEATURES OF INTEREST
Alpha (α) Boötis *Arcturus*	-0.04	A red giant that sits at a distance of 75 light-years from the Solar System.
Mu (μ) Boötis *Alkalurops*	4.3	A binocular double with a mag. 7.0 companion (itself two very close stars). The stars are yellow and bluish.
Nu (ν) Boötis	5.0	A naked-eye double with a mag. 5.0 companion 10' 30" away. The stars are orange and white.

ARCTURUS 👁

Now in old age, Arcturus is running out of fuel, and internal processes are turning it into a red giant. Currently, its diameter is 25 times that of the Sun, or about 24.9 million miles (40 million km), although that is set to increase by a factor of 10. When the Sun goes through the same process, its atmosphere will expand to beyond the extent of the Earth's orbit, thus dooming it and the rest of the Solar System's inner planets.

CES (Com)

2nd

0°N–56°S

SKY MAP: COMA BERENICES

To the south of Gamma (γ) Comae Berenices is the open cluster Mel 111, which is visible to the eye.

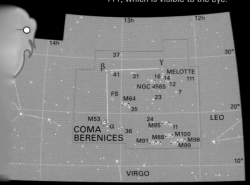

OBSERVING COMA BERENICES

MAJOR STARS	MAGNITUDE	FEATURES OF INTEREST
Alpha (α) Comae Berenices *Diadem*	4.3	A binary star 65 light-years away.
24 Comae Berenices	5.0	A great double star with a mag. 6.7 companion 20" away. The stars are yellow and blue.

NOTABLE OBJECTS	MAGNITUDE	FEATURES OF INTEREST
M53	7.6	A globular cluster with a slight oval appearance in a telescope.
Melotte (Mel) 111, the Coma Star Cluster	2.7	An open star cluster with around 45 starry members.

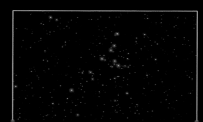

MEL 111 👁

The Coma Star Cluster, also known as Mel 111, looks great in binoculars. Classically, this was the tuft at the end of Leo's tail until Coma Berenices was created in 1602.

Berenice of **Egypt** cut off her
as a tribute to the **gods** after
shand's return from **battle**

VIRGO

Virgo, the Maiden, is the second-largest constellation in the night sky, and is home to the many thousands of galaxies of the eponymous Virgo Cluster.

The most obvious star in this big, sprawling constellation is Alpha (α) Virginis, or Spica. It is a blue giant star, sitting 220 light-years away from us and positioned just below the ecliptic line. This location means Spica can be covered, or occulted, by the Moon and sometimes the planets—although the latter is extremely rare. Until recent advances in technology, these occultations were one way of calculating orbital information about the Moon. Spica is one of only four bright stars that can be occulted in this way, the others being Antares, in Scorpius; Regulus, in Leo; and Aldebaran, in Taurus. One of Virgo's most prominent objects is the Sombrero Galaxy, but the constellation is also home to the Virgo Cluster, the nearest large galaxy cluster to Earth, 50 million light-years distant. Its brightest members are giant elliptical galaxies, notably M87 and M49.

Latin for "ear of wheat," **Spica** marks the **bounty** held in the left hand of the **Maiden**.

SPICA

M104, THE SOMBRERO GALAXY ✦

M87 ✦

At the heart of the Virgo Cluster, the galaxy M87 has a diameter roughly equivalent to that of the Milky Way, and through a small telescope looks like a roundish blob with a noticeably brighter center. Larger telescopes reveal an unusual straight-line feature, which is actually a jet of material streaming out of the galaxy. Investigations have shown that M87 is a fascinating object—the jet is some 5,000 light-years long and is moving at nearly the speed of light. Studies reveal that this feature is probably caused by material being blasted out of the galaxy by a supermassive black hole that sits at its core. M87 is also a strong emitter of radio waves and X-rays, which are also indicators of a black hole.

JET-PROPELLED GALAXY
Emitting an enormous jet of material (left), M87 is a massive elliptical galaxy that is surrounded by starlike globular clusters (above).

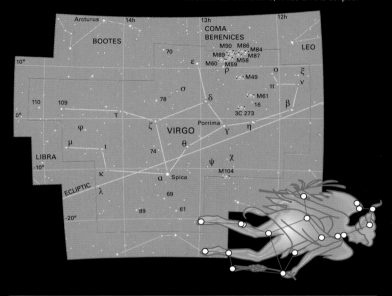

M49 ✕

PORRIMA

WIDE VIEW 👁
Despite its plethora of galaxies, the only object of note to the naked eye in this large constellation is Spica, the 15th-brightest star in the night sky.

VIRGO (Vir)

WIDTH / DEPTH	✋✋✋✋ / ✋✋
SIZE RANKING	**2nd**
FULLY VISIBLE	**67°N–75°S**

SKY MAP
Virgo sits across the celestial equator, so it can be seen equally well from the northern and southern hemispheres. As one of the zodiacal constellations, it lies on the ecliptic.

OBSERVING VIRGO

MAJOR STARS	MAGNITUDE	FEATURES
Alpha (α) Virginis *Spica*	1.0	A blue star, part of a binary system so close that both stars are ellipsoidal in shape due to gravity.
Gamma (γ) Virginis *Porrima*	3.50	A beautiful double star with a companion of identical magnitude. The apparent distance between them varies (see below). They are creamy white in color.

NOTABLE OBJECTS	MAGNITUDE	FEATURES
M49	8.4	An elliptical galaxy 60 million light-years distant.
M87	9.5	A giant elliptical galaxy also known as Virgo A.
M104, the Sombrero Galaxy	9.0	A spiral galaxy with a thick ring of dust giving it a Mexican-hat-like appearance.

M104, the Sombrero Galaxy ✕

This spectacular galaxy has plenty of history: studies in the early 1900s showed that M104 was moving away from the Milky Way—leading to the proof that the Universe was expanding, and so contributing to the formulation of the Big Bang theory. You can view this historically significant galaxy with binoculars, but will need a good-size telescope to see the bulbous core and dark rim of dust that gives it its striking, sombrero-like appearance. From Earth, we see M104 from six degrees above its equatorial plane, meaning that we get a clear view of its core as well as its spiral arms. The Sombrero Galaxy is not part of the Virgo galaxy cluster, lying around only two-thirds of the distance from Earth to M87 (see panel, opposite).

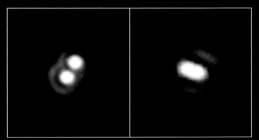

PORRIMA ✕
The double star Gamma (γ) Virginis, or Porrima, could last be seen separately in a small telescope (far left) in the early 1990s. The two stars are currently orbiting each other very closely (left), making it difficult to separate them without a large telescope.

CORONA BOREALIS AND SERPENS CAPUT

The curving group that is Corona Borealis, the Northern Crown, sits directly to the north of Serpens Caput, the Serpent's Head, in this quiet region of the northern night sky.

The best way of finding Serpens Caput is to use Corona Borealis, as the Crown's seven main stars form a lovely semi-circular chain with the ends pointing slightly off-north. There is only one star in Corona Borealis with any real brightness and that is Alpha (α) Coronae Borealis, known in Arabic as Alphecca—the Bright Star of the Broken Ring of Stars—and in Latin as Gemma, which translates more simply as Jewel. No deep-sky objects are visible for most amateur astronomers, but the constellation does have some fine variable stars and a good double for viewing in binoculars, Nu (ν) Corona Borealis. Move directly south of Corona Borealis to enter Serpens Caput territory. The main feature here may only be a single deep-sky object in its southwestern corner, the globular cluster M5, but it really repays observation. A faint misty smudge to the eye in a perfectly dark sky, it is an easy target for binoculars and a fine sight in a small telescope.

R Coronae Borealis 🔭

This most unusual variable star has the nickname of "the Reverse Nova." While a normal nova star builds up material until there is a bright outburst, in this one there is a build-up of material over several years that causes the star to fade away. The result is a variable with an exceptional range in magnitude, normally around mag. 5.8 but ending up somewhere near 14.8. What is happening here is that the brightness of the star gradually reduces as dark carbon dust, like soot, builds up in its atmosphere. When there is too much for the star to cope with, it puffs the soot away, and the brightness gradually returns. This is a star for binoculars when at its brightest and a telescope when at its faintest.

CORONA BOREALIS

ALPHECCA

WIDE VIEW 👁
The curving horseshoe pattern of seven stars that make up Corona Borealis can be appreciated best from darker skies, as there is only one really bright star.

SERPENS CAPUT

UNUKALHAI

M5

WIDE VIEW 👁
The constellation only has one feature of note, and that is the three stars that together form a neat triangle representing the head of the Serpent.

CORONA BOREALIS (CrB)

WIDTH / DEPTH	🤏 / 🤏
SIZE RANKING	**73rd**
FULLY VISIBLE	**90°N–50°S**

SKY MAP: CORONA BOREALIS

The Crown sits snugly between the Keystone of stars that identify Hercules and the kite-shape of Boötes, the Herdsman.

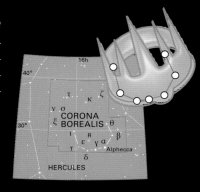

NU CORONAE BOREALIS 🔭

Right on the Crown's western border, the stars ν¹ and ν² form a wonderful double for binoculars, with both showing a distinct reddish color.

OBSERVING CORONA BOREALIS

MAJOR STARS	MAGNITUDE	FEATURES OF INTEREST
Alpha (α) Coronae Borealis *Alphecca*	2.2	A bright star lying at a distance of 75 light-years.
Nu-1 (ν¹) Coronae Borealis	5.2	A wide double with ν², mag. 5.4, as its companion 6' away. Both are red.
R Coronae Borealis	5.8	An irregular variable star that dims to around mag. 14.8.
T Coronae Borealis *Blaze Star*	10.8	A recurring nova with a period around 9,000 days over which the magnitude soars from 10.8 to 2.0.

The crown of the Minoan princess **Ariadne** was hurled into the sky by the lovestruck god **Dionysus**.

SERPENS (Ser)

WIDTH / DEPTH	🤚🤚🤏 / 🤚🤚
SIZE RANKING	**23rd**
FULLY VISIBLE	**74°N–64°S**

SKY MAP: SERPENS CAPUT

The constellation represents the upper half of the serpent that is held by Ophiuchus, the Serpent Bearer (see pp.100–101).

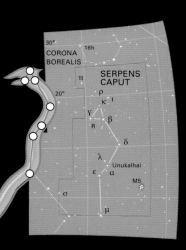

OBSERVING SERPENS CAPUT

MAJOR STARS	MAGNITUDE	FEATURES OF INTEREST
Alpha (α) Serpentis *Unukalhai*	4.6	A red giant with a diameter 15 times larger than the Sun.
R Serpentis	5.2	A long-period Mira-type variable star with a period of 356 days that sees it dim to around mag. 13.8.

NOTABLE OBJECTS	MAGNITUDE	FEATURES OF INTEREST
M5	5.6	A very fine globular cluster easily viewed in binoculars or a small telescope.

M5 🖥

M5 is one of the largest and oldest globular clusters known, with a width of 165 light-years and an estimated age of around 13 billion years. It sits 24,500 light-years away and is believed to contain at least 100,000 stars. The finest view of M5 was obtained when it was viewed from outside the Earth's atmosphere by the Hubble Space Telescope. The result was this spectacular image.

OPHIUCHUS AND SERPENS CAUDA

These two constellations can often be overlooked, yet they contain objects that put a whole new perspective on the workings of the Universe: the Eagle Nebula and Barnard's Star.

These constellations are placed together not only because they are adjacent to one another in the sky, but also because they are connected in mythology. Ophiuchus is the Serpent Bearer, although this is only reflected in the original name of the constellation, which was Ophiuchus vel Serpentarius. In classical art, Ophiuchus holds the Serpent in both hands, with the head, Serpens Caput, in his left and the tail, Serpens Cauda, in his right. Focusing on Serpens Cauda, the area neighboring Ophiuchus, stargazers can find one of the great star-making areas, known as the Eagle Nebula, M16. This area has been spectacularly photographed by the Hubble Space Telescope. Ophiuchus is the 11th-largest constellation, and it contains a large, empty-looking patch of sky. On closer inspection—for example, through binoculars—the area reveals several good globular and open star clusters, such as M10 and M12. Ophiuchus is also home to the amazing Barnard's Star, which you can observe as it travels through space over a period of a few years.

Barnard's Star 🔭
The red dwarf Barnard's Star is the fourth-closest star to the Sun, at a distance of 5.9 light-years, yet is so dim that it only registers at magnitude 9.5. It was discovered by the astronomer Edward Barnard in 1916, who measured its flight across the night sky at an incredible rate of 10 arcseconds (also written as 10") per year. This equates to it covering the distance of one width of the full Moon in just 188 years. Astronomically, this is called "proper motion," and although Barnard's Star is the fastest, it is not alone—all stars have proper motion, yet most are too far away, or move too slowly, to be noticeable over a human lifetime. This "flying star" is currently moving north of a line between Beta and 66 Ophiuchi.

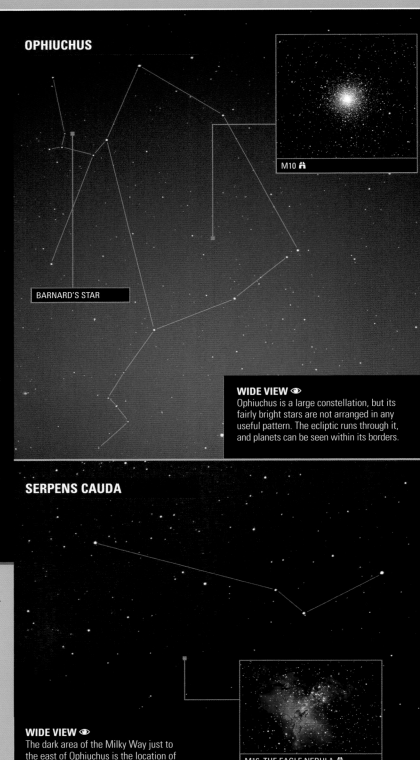

OPHIUCHUS

BARNARD'S STAR

M10 🔭

WIDE VIEW 👁
Ophiuchus is a large constellation, but its fairly bright stars are not arranged in any useful pattern. The ecliptic runs through it, and planets can be seen within its borders.

SERPENS CAUDA

WIDE VIEW 👁
The dark area of the Milky Way just to the east of Ophiuchus is the location of Serpens Cauda. M16, the Eagle Nebula, lies below the main star figure.

M16, THE EAGLE NEBULA 🔭

...CHUS (Oph)

...DEPTH 🖐🖐🖐 / 🖐🖐

...NKING **11th**

...SIBLE **59°N–75°S**

...AP:
...CHUS
...as of Ophiuchus
...to Hercules and
...s are the most
...ting, as they sit
...the Milky Way.

OBSERVING OPHIUCHUS

MAJOR STARS	MAGNITUDE	FEATURES OF INTEREST
Alpha (α) Ophiuchi *Rasalhague*	2.1	The brightest star in Ophiuchus, about 30 times more luminous than the Sun.
Rho (ρ) Ophiuchi	5.1	A double star with a close mag. 5.7 companion just 5″ away; a great treat in a small telescope.
TYC 425-2502-1 *Barnard's Star*	9.6	A close (6-light-year distant) red dwarf, the fastest-moving star in the sky.

NOTABLE OBJECTS	MAGNITUDE	FEATURES OF INTEREST
M10	6.6	A globular cluster, observable in binoculars, close to M12.
M12	6.6	A globular cluster with a larger, less compact appearance than M10.
IC 4665	4.2	A large, loose open star cluster of around 30 stars just above Beta (β) Ophiuchi, or Cebalrai.
NGC 6633	4.6	An open star cluster; a fine group of 30 stars on the border with Serpens Cauda.

Ophiuchus, the **Serpent Bearer**, **straddles** the celestial **equator,** and holds the **"snake"** of Serpens Caput and Serpens Cauda in his hands.

...PENS (Ser)

...H / DEPTH 🖐🖐🖐 / 🖐🖐

...RANKING **23rd**

...Y VISIBLE **74°N–64°S**

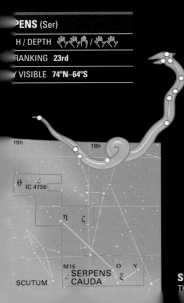

SKY MAP: SERPENS CAUDA
The Milky Way gives Serpens

THE PILLARS OF CREATION 🔭
M16, the Eagle Nebula, includes a cluster of stars visible in binoculars. It is also home to the amazing columns of dust and gas known as the Pillars of Creation.

OBSERVING SERPENS CAUDA

MAJOR STARS	MAGNITUDE	FEATURES OF INTEREST
Theta (θ) Serpentis *Alya*	4.6	A good double for telescope viewing with a mag. 5.0 companion, 20″ away. Both stars are white.
Xi (ξ) Serpentis *Nehushtan*	3.5	A double star with a mag. 6.0 companion lying directly south; both are just visible to the naked eye.

NOTABLE OBJECTS	MAGNITUDE	FEATURES OF INTEREST
M16, the Eagle Nebula	6.0	A fascinating bright nebula sitting 7,000 light-years away.
IC 4756	4.6	An open star cluster of around 80 stars.

LYRA AND HERCULES

Each of these constellations contains the finest observable example of one type of deep-sky object: Hercules is home to the Great Globular, M13, while Lyra houses the planetary Ring Nebula, M57.

Sitting side-by-side in the sky, Lyra, the Lyre, and Hercules, the Strong Man, are both fine-looking constellations, with bright stars arranged in recognizable shapes. The four central stars of Hercules form what is known as the Keystone. The northern star of this asterism is Eta (η) Herculis, Sofian, and the stunning globular star cluster M13—the Great Globular—lies one-third of the journey south from here toward Zeta (ζ) Herculis, Ruticulus. M13 is just visible to the naked eye, but is a marvelous sight with binoculars or a small telescope. Close by is the constellation of Lyra, containing Alpha (α) Lyrae, or Vega, the fifth-brightest star in the sky and one of the stars of the famous Summer Triangle trio (see pp.108–109). To the north of Vega is the quadruple star Epsilon (ε) Lyrae; to see this with the naked eye as a double is supposedly a proof of good eyesight. A telescope—the bigger the better—is essential to view Lyra's extraordinary Ring Nebula, a planetary nebula that resembles a ring of smoke.

M13, the Great Globular cluster 👁

The Great Globular, M13, is a giant, spherical island of perhaps 250,000 stars, and is around 12 million years old. It lies about 25,000 light-years away from Earth, on the edge of the Keystone of Hercules, and is the brightest globular cluster in the northern hemisphere. Even so, it is still barely visible to the naked eye, looking like a hazy star, and in urban areas it can often only be observed with the aid of binoculars, in which it appears as half the apparent width of the full Moon. Through a small telescope M13 breaks up into countless starry points. A similar sight for southern-hemisphere skywatchers is the gigantic globular Omega Centauri, the largest in the Milky Way, in the constellation Centaurus (see p.142).

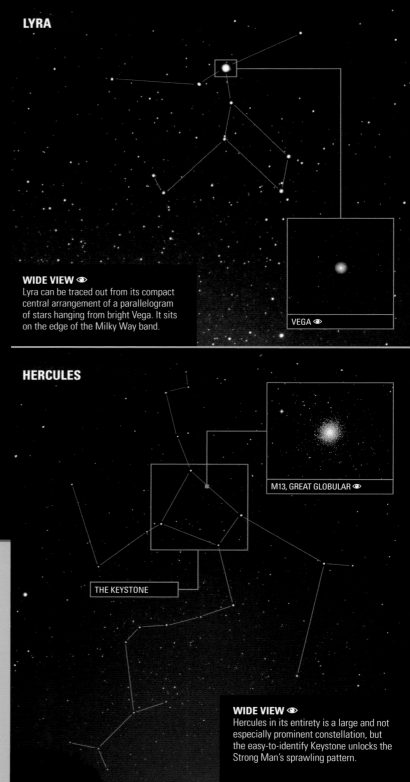

LYRA

WIDE VIEW 👁
Lyra can be traced out from its compact central arrangement of a parallelogram of stars hanging from bright Vega. It sits on the edge of the Milky Way band.

VEGA 👁

HERCULES

M13, GREAT GLOBULAR 👁

THE KEYSTONE

WIDE VIEW 👁
Hercules in its entirety is a large and not especially prominent constellation, but the easy-to-identify Keystone unlocks the Strong Man's sprawling pattern.

A (Lyr)

TH / DEPTH	🖐/🖐🖐
E RANKING	**52nd**
LY VISIBLE	**90°N–42°S**

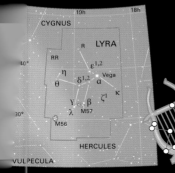

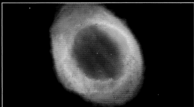

M57, THE RING NEBULA ✈

Through a small telescope, the famous Ring Nebula in Lyra appears as a bright oval; through a larger telescope, the "smoke-ring" shape, formed by the expanding sphere of cast-off debris from a former star, comes into view.

SKY MAP: LYRA

The lyre shape is not easy to imagine from this constellation's pattern, which is most easily found by spotting Vega, its brightest star.

The mythical musician **Orpheus** played the **lyre** on his journey into the **Underworld**.

OBSERVING LYRA

MAJOR STARS	MAGNITUDE	FEATURES OF INTEREST
Alpha (α) Lyrae *Vega*	0.0	A bright, bluish star that sits 26 light-years away.
Beta (β) Lyrae *Sheliak*	3.5	A variable star ranging between mag. 3.3 and 4.4 over 13 days.
Delta-1 (δ¹) Lyrae	5.6	A wide double star visible to the naked eye.
Epsilon (ε) Lyrae	4.7	The famous "Double Double" of Lyra: each of the two bright stars is also itself a double.
Zeta (ζ) Lyrae	4.3	A double star observable in binoculars with a mag. 5.9 companion 45" away. The star colors are topaz and greenish.

NOTABLE OBJECTS	MAGNITUDE	FEATURES OF INTEREST
M57, the Ring Nebula	9.0	A planetary nebula seen as a ghostly oval through a small telescope.

HERCULES (Her)

WIDTH / DEPTH	🖐🖐🖐 / 🖐🖐🖐
SIZE RANKING	**5th**
FULLY VISIBLE	**90°N–38°S**

OBSERVING HERCULES

MAJOR STARS	MAGNITUDE	FEATURES OF INTEREST
Alpha (α) Herculis *Rasalgethi*	3.1	A very red variable star which dims erratically to mag. 4.1. It has a diameter 400 times larger than the Sun.
Kappa (κ) Herculis	5.3	A double star for telescope viewing, with a mag. 6.5 companion 30" away. The stars are both orange.

NOTABLE OBJECTS	MAGNITUDE	FEATURES OF INTEREST
M13, the Great Globular	5.7	A fantastic globular cluster lying around 25,200 light-years from Earth.
M92	6.5	A globular cluster, not as big or as bright as M13, but still an impressive sight.

SKY MAP: HERCULES

Hercules is bordered by nine constellations, and so its stars can be used to map out a large part of the night sky.

ABELL 2151 CLUSTER ⚖

With a powerful telescope, this is a marvelous sight. Each of the spirals and weirdly-shaped irregular objects is a galaxy, all interacting with each other in an early stage of development

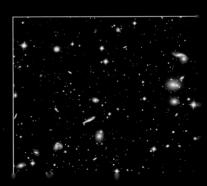

CYGNUS

The wonderful, chance arrangement of the five main stars in Cygnus, the Swan, leads the constellation to be known familiarly as the Northern Cross.

The brightest star in Cygnus is Deneb, Alpha (α) Cygni—the tail of the Swan. Its head is marked by Albireo, or Beta (β) Cygni, one of the finest double stars in the sky. The outstretched wings of Cygnus extend on each side from the three stars that form the bar of the "cross." Halfway between the tail and the right wing-tip is 61 Cygni—not only a double star, but also a moving one, whose progress can be tracked over the years. A broad telescopic view of Cygnus hints at myriad stars, clusters, and nebulae awaiting closer inspection. Just to the right of the tail, NGC 7000 can be glimpsed with the naked eye from dark locations. Its distinctive shape, resembling the coastline of Mexico and the USA, has naturally led to it being called the North America Nebula.

Centered on the Milky Way, Cygnus is one of the northern hemisphere's finest constellations.

Deneb and the Summer Triangle 👁
Deneb translates from Arabic as "tail." Many star names are just literal translations of what feature they represent in the constellation's classical figure, and as there are a lot of birds and animals peppering the heavens, there are many stars with variations on the name "Deneb." Deneb, Altair in Aquila, and Vega in Lyra form the Summer Triangle (see pp.106–107), one of the most famous arrangements of stars in the northern hemisphere. Deneb seems to us less bright than the other two; however, while Altair and Vega, at 16 and 25 light-years respectively, are relatively close to Earth, Deneb is an incredible 1,500 light-years away. The luminosity of this blue-white supergiant has been estimated at 60,000 times that of the Sun.

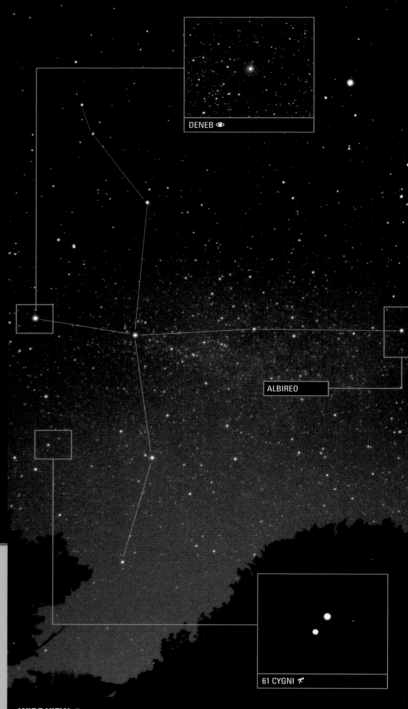

DENEB 👁

ALBIREO

61 CYGNI ⚹

WIDE VIEW 👁
The naked-eye view of Cygnus is spectacular as there is a dark rift in the Milky Way—the Cygnus Rift or Northern Coalsack—over which the Swan flies.

CYGNUS (Cyg)

WIDTH / DEPTH	
SIZE RANKING	16th
FULLY VISIBLE	90°N–28°S

SKY MAP

Cygnus is one of the easiest constellations to visualize, as a swan flying southward along the Milky Way.

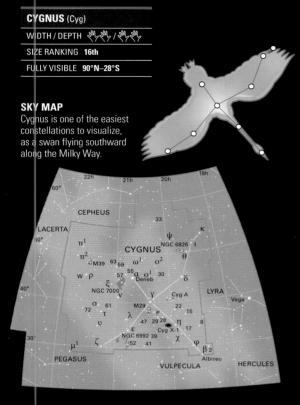

OBSERVING CYGNUS

MAJOR STARS	MAGNITUDE	FEATURES
Alpha (α) Cygni *Deneb*	1.3	A huge, distant star with a diameter that is likely to be 300 times that of the Sun.
Beta (β) Cygni *Albireo*	3.2	A beautiful double star with a mag. 5.1 companion 35″ away. The stars are golden and blue in color.
Omicron-1 (o¹) Cygni (31 Cygni)	3.8	A good double star for binoculars, with a mag. 4.8 companion 1′ 30″ away. The stars are orange and turquoise in color.
61 Cygni	5.2	A double star with a mag. 6.0 companion 30″ away. The stars are orange in color.

NOTABLE OBJECTS	MAGNITUDE	FEATURES
M39	4.6	An open star cluster of around 30 stars some 825 light-years away.
NGC 6960, 6992, & 6995, the Veil Nebula	7.0	Wispy remains, visible in a large telescope, that are the debris of an ancient supernova explosion.
NGC 7000, the North America Nebula	4.0	A bright nebula four times the size of the full Moon, although its shape is not visible to the naked eye.

VEIL NEBULA

These gaseous remains are spreading out from a center once occupied by a star that exploded. All in all the phenomenon is known as the Cygnus Loop, but those portions of it that are visible are, because of their almost ethereal, wispy appearance, called the Veil Nebula. The extent of their spread suggests that the supernova explosion took place between 5,000 and 15,000 years ago.

ALBIREO

In a small telescope the double stars of Albireo, a golden giant and a blue dwarf, are widely spaced. The larger star has its own dwarf companion, too close and small to be seen.

NGC 7000, the North America Nebula

Just next to Deneb sits a slightly brighter patch of the Milky Way, which is the famous North America Nebula, NGC 7000. Like all nebulae, this is simply a cloud of gas, and maybe dust, that occupies an area of space. Nebulae are similar to clouds in the sky, in that their appearance depends to a great extent on how they are lit. Some nebulae are currently forming stars and their light can make the nebula itself glow; others are dark and are only seen because they are silhouetted against any brighter background. NGC 7000 is a fine example of a bright nebula, visible because it is being lit by the powerful light of the nearby giant star Deneb.

NGC 7000

The resemblance of this nebula to a map of North America is to a great extent suggested by a dark area reminiscent of the Gulf of Mexico (center left).

STARHOPPING FROM THE SUMMER TRIANGLE

The brightest three stars in the evening skies of summer and fall in the north make a simple triangular signpost for locating nearby constellations.

The Summer Triangle is not a constellation, but an asterism. Its three points are Altair in the constellation of Aquila, Deneb in Cygnus, and Vega in Lyra. The Summer Triangle lies almost overhead at mid-northern latitudes during the summer months.

1 TO HERCULES

Locate the bright star Deneb in Cygnus, the Swan, which marks one of the points of the Summer Triangle. From here, draw an imaginary line along

the shortest side of the Triangle toward the star Vega. Extend this line further, continuing it for about the same distance in the sky as that from Deneb to Vega, and your gaze should fall upon the constellation of Hercules (see pp.102–103), with its distinctive four-star asterism, the "Keystone," off to one side.

HERCULES

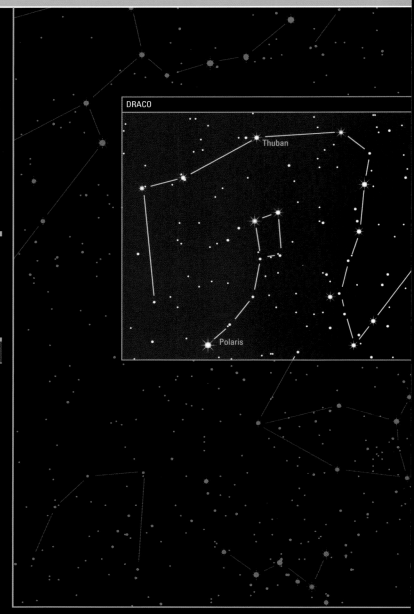

DRACO

Thuban

Polaris

2 TO SAGITTARIUS

Starting again at Deneb—the faintest of the three stars in the Summer Triangle—trace an imaginary line to Altair, the leading star of Aquila, the Eagle. Note that just before this star is reached, the line travels through the small constellation of Sagitta, the Arrow. Continue this line approximately the same distance as from Deneb to Altair, and you will arrive at the constellation of Sagittarius, with its distinctive Teapot asterism.

OPHIUCHUS

3 TO OPHIUCHUS

Ophiuchus can be a very difficult constellation to locate, but thankfully three stars near Altair in the constellation Aquila will aid a stargaze here. Start at the "beak" of the eagle, at Theta (θ) Aquilae, and draw an imaginary line through the central star, Bezek (Eta (η) Aquilae). This straight line will also pass through Delta (δ) Aquilae. Extend the line out of the constellation of Aquila; the next fairly bright star you reach is Basalhague, in Ophiuchus.

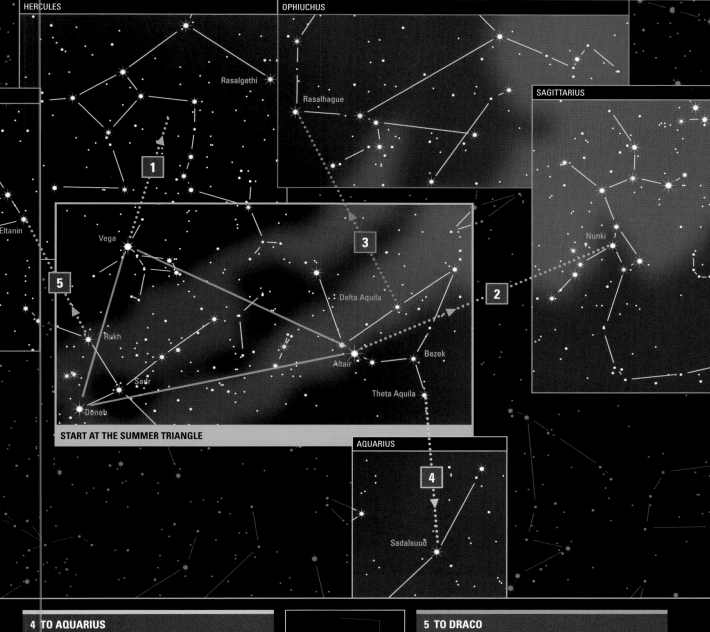

HERCULES

Rasalgethi

OPHIUCHUS

Rasalhague

SAGITTARIUS

Nunki

Eltanin

1

Vega

5

3

Delta Aquila

2

Rukh

Sadr

Altair

Bezek

Theta Aquila

Deneb

START AT THE SUMMER TRIANGLE

AQUARIUS

4

Sadalsuud

4 TO AQUARIUS

Using the same three locator stars for Ophiuchus, but in the reverse direction, Aquila can also be a signpost to the constellation of Aquarius, the Water Carrier. Start at the star Delta (δ) Aquilae, found near Altair, the main star of Aquila, the Eagle. Extend a line through Bezek (Eta (η) Aquilae) and out beyond Theta (θ) Aquilae; you will very soon reach the star Sadalsuud in Aquarius, which is the Water Carrier's left shoulder.

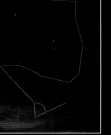

DRACO

5 TO DRACO

Deneb lies on the Swan's "tail" in the prominent constellation of Cygnus. Find the central star of the Swan's body, Gamma (γ) Cygni or Sadr, and draw a line from here to the wing star Delta (δ) Cygni, Rukh. Extend the line further and you will meet two bright stars in the constellation of Draco, the Dragon. Adjacent to these are two fainter stars; together, these four stars make up the lozenge-shaped head of the Dragon.

AQUILA

Aquila, the Eagle, sits across the Milky Way, with its main star Alpha (α) Aquilae, better known as Altair, forming the southern point of the famously bright Summer Triangle pattern.

Aquila contains a good number of bright stars, all arranged in a fairly easy-to-identify pattern, and also shares borders with no fewer than nine other constellations. However, it is the bright leading star Altair that really helps to locate the Eagle, not least because it forms, with Vega in Lyra and Deneb in Cygnus, the Summer Triangle, a northern-hemisphere asterism so bright that it stands out even in the most light-polluted locations. However, in a dark sky, Aquila itself soars magnificently against the background star fields of the Milky Way. Look out for Eta (η) Aquilae, a Cepheid variable star which changes brightness in just over a week, and the nebulae B142 and B143—two large, dark patches in the Milky Way, best seen with binoculars.

> The **Eagle soars** magnificently against the background star fields of the **Milky Way**.

NGC 6709 ⛗

On Aquila's edge, close to the border with neighboring Hercules, the open star cluster NGC 6709 sits within the Milky Way where the scene is already awash with stars. Nevertheless, the cluster's concentration of around 100 stars in a triangular formation, with a bright star at each corner, makes it relatively easy to identify with binoculars. With objects such as this, where individual stars are somewhat faint, the aperture size of binoculars can really make a difference. With apertures up to 60mm, NGC 6709 is a misty patch, while 70mm or more will resolve the individual stars. With these heavy large-aperture binoculars, a tripod is usually necessary for support and steady viewing.

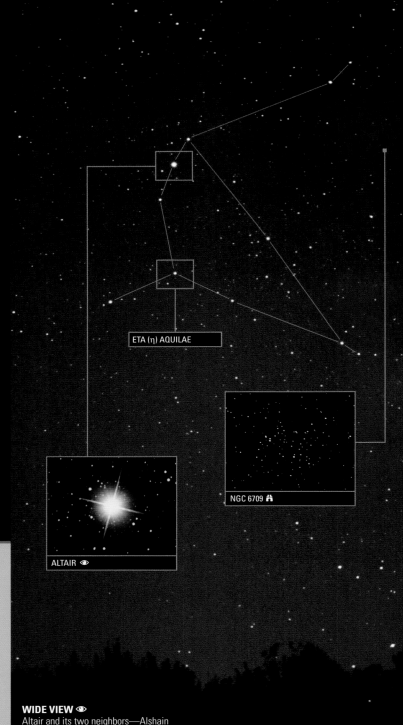

ETA (η) AQUILAE

NGC 6709 ⛗

ALTAIR 👁

WIDE VIEW 👁
Altair and its two neighbors—Alshain and the orangey Tarazed—form a distinctive gentle curve from which the remaining stars in Aquila can be located.

AQUILA (Aql)

WIDTH / DEPTH	
SIZE RANKING	**22nd**
FULLY VISIBLE	**78°N–71°S**

SKY MAP

The most common visualization of Aquila places the eagle slightly off-center, with a broad, sweeping wingspan. The classical representation (right) often places lightning bolts in Aquila's beak, a reference to the eagle that carried these weapons for the Greek god Zeus.

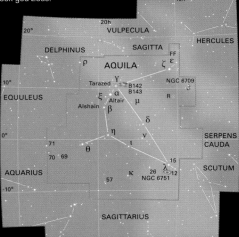

OBSERVING AQUILA

MAJOR STARS	MAGNITUDE	FEATURES
Alpha (α) Aquilae *Altair*	0.8	The 12th-brightest star in the sky and just 16.8 light-years away from us.
Eta (η) Aquilae *Bezek*	3.9	A Cepheid variable star (see p.90) with a mag. range from 3.5 to 4.3 over 7 days.

NOTABLE OBJECTS	MAGNITUDE	FEATURES
NGC 6709	7.6	An open star cluster of around 100 stars sitting about 3,000 light-years away.
B142 & B143, the "E" Nebula	—	A dark nebula consisting of unlit clouds within the Milky Way.

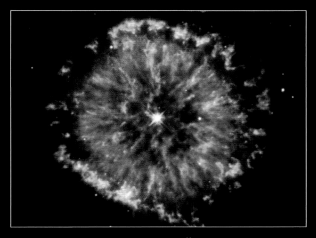

NGC 6751, THE GLOWING EYE NEBULA

This eerie Hubble image captures a planetary nebula in Aquila 6,500 light-years away that is only visible through powerful telescopes. It was created using advanced imaging techniques in which color filters were used to isolate gaseous emissions at different temperatures, blue representing the hottest.

B142 and B143

Generally, it is the bright objects in the Universe, either puncturing the darkness or filling it with a bright haze of whatever construction, that draw our attention. However, there are also intriguing dark phenomena that deserve the stargazer's attention. The dark nebulae B142 and its neighbor B143 are such dark objects. Their combined shapes make up the "E" Nebula, named by the astronomer Edward Barnard, who cataloged dark nebulae in the early 20th century. In the night sky it looks as though the stars have avoided this letter-E-shaped area. The effect is not caused by the absence of stars at all, but by great dark clouds of dust and gas that are simply obscuring our view of the stars behind.

THE "E" NEBULA

Comprising two dark nebulae, this dark shape covers an area of sky roughly equivalent to the full Moon, and has been estimated to be some 2,000 light-years away.

VULPECULA

What Vulpecula, the Little Fox, lacks in any sort of memorable star pattern is made up for by the quality of its features. This is mainly because it sits across a bright section of our galaxy.

Vulpecula can easily be totally overlooked, because it is a small and faint constellation that sits in a wonderfully busy section of the Milky Way, sandwiched between the dominant constellations of Cygnus, Lyra, and Aquila. The Fox does not have any bright stars, so it is not surprising that many non-stargazers have never heard of it. Despite its relative obscurity, however, Vulpecula is home to two unmissable celestial objects easily seen in binoculars: the Coathanger, more formally known as Brocchi's Cluster; and M27, the Dumbbell Nebula. Vulpecula's leading star, Alpha (α) Vulpeculae or Anser, "the Goose" (see below), is a 4th-magnitude red giant. Its nearby, unrelated 6th-magnitude companion is visible through binoculars.

This **faint** group of stars is home to the **Coathanger** and the **Dumbbell** Nebula.

A modern constellation
The common name of Alpha Vulpeculae is Anser. This comes from the constellation's original name, Vulpecula cum Anser (the little fox with the goose). This is a modern constellation, "invented" in the 17th century by a remarkable Polish astronomer, Johannes Hevelius. He proposed the creation of 10 new constellations, seven of which were officially adopted. Others include Canes Venatici and Scutum, which Hevelius originally designated Scutum Sobiescanum, in honor of the Polish king, John III Sobieski. None are particularly striking to the naked eye, because Hevelius had to use areas of the sky with few bright stars or notable star patterns—the best were already used in the major constellations.

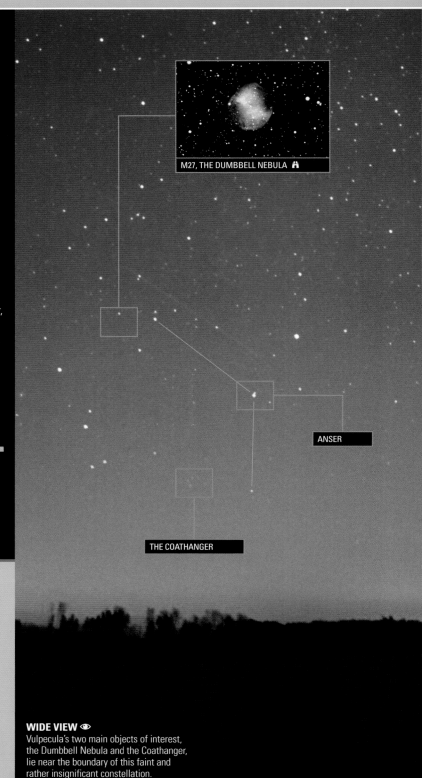

M27, THE DUMBBELL NEBULA

ANSER

THE COATHANGER

WIDE VIEW 👁
Vulpecula's two main objects of interest, the Dumbbell Nebula and the Coathanger, lie near the boundary of this faint and rather insignificant constellation.

VULPECULA (Vul)

WIDTH / DEPTH	〈hand symbols〉
SIZE RANKING	55th
FULLY VISIBLE	90°N–61°S

SKY MAP
The constellations of Pegasus and Hercules can help to locate Vulpecula, as can the bright star Vega, part of the Summer Triangle.

OBSERVING VULPECULA

MAJOR STARS	MAGNITUDE	FEATURES
Alpha (α) Vulpeculae *Anser*	4.4	A double star with a mag. 5.8 companion 6' 50" away. The stars are red and orange.

NOTABLE OBJECTS	MAGNITUDE	FEATURES
M27, the Dumbbell Nebula	7.6	A planetary nebula that looks great when viewed with a small telescope.
Collinder 399, the Coathanger	3.6	An asterism best seen with binoculars.

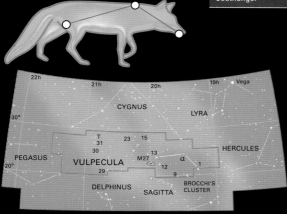

THE COATHANGER 𝍫
Also known as Brocchi's Cluster, or Collinder (CR) 399, the Coathanger is so-named due to its amazing arrangement of ten stars. To the naked eye it appears as a small misty smudge in dark skies, but binoculars easily reveal its unique shape. Initially thought to be a real cluster, recent studies seem to indicate that it is just a chance alignment of stars.

COLORFUL IMAGING
CCD imaging (left) adds color to the double-lobed nebula, M87, which is formed from gassy clouds ionized by radiation (seen close-up in the Hubble Space Telescope image above).

M27, the Dumbbell Nebula 𝍫
This nebula's twin-lobed shape is reminiscent of a pair of dumbbells. It sits to the south of the star 13 Vulpeculae, which forms the lower-central point of a great M-shaped group of stars. Coupled with the surrounding buzz of the Milky Way, this makes M27 a wonderful object to observe in a fascinating star field. M27 can be seen with binoculars, appearing as a round patch a quarter of the size of the full Moon, but a telescope is necessary to view detail. It is a planetary nebula, around 1,000 light-years distant from Earth, and was created when a dying star threw off its atmosphere, which then expanded into space.

PEGASUS AND PISCES

Pegasus, the Winged Horse, is the adjacent constellation to the much fainter Pisces, the Fishes. The greatest asset of Pegasus is its Great Square of stars, which can help you to find nearby Pisces easily.

The constellations of Pegasus and Pisces are part of a large and quiet area of the night sky. There are not many bright stars around here, nor does the Milky Way, with its astronomical treasures, flow through either of these constellations. What Pegasus does offer, however, is a pattern of four stars, fairly regularly spaced, that forms the asterism known as the Great Square of Pegasus. This prominent feature acts as a tremendous "signpost," helping to find nearby stars and constellations. Below the side of the Square, between Algenib and Markab, is a faint asterism of seven stars known as the Circlet, which represents one of the fish in Pisces. Due to the sparseness of bright stars in this region of the sky, this little oval-looking group of fourth- and fifth- magnitude stars is not difficult to find. The rest of the constellation follows a faint line of stars and then changes direction at the leading star, Alrescha, Alpha (α) Piscium, where it zigzags up to the head of the second "fish."

M15 📷

The globular cluster M15 is a tremendous-looking, distant, spherical island of stars, one of the densest in the Milky Way. An estimated age of 13.2 billion years makes it one of the oldest globular clusters in the Universe. Indeed, if this estimate is correct, then M15 was created not that long after the Universe itself came into being. M15 is 175 light-years wide, and looks like a slightly more compact version of the Great Globular, M13, in Hercules. It is found by starhopping down Pegasus's head, from Theta (θ) Pegasi on his forelock in a straight line to Epsilon (ε) Pegasi, or Enif, "the nose," and beyond. It is easily observable in binoculars, but a telescope is needed to distinguish the individual stars.

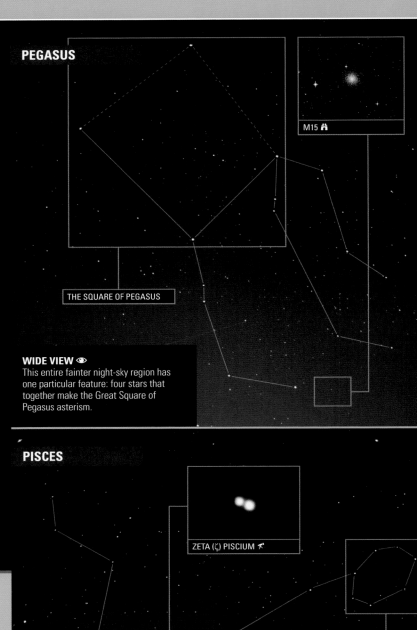

PEGASUS

M15 📷

THE SQUARE OF PEGASUS

WIDE VIEW 👁
This entire fainter night-sky region has one particular feature: four stars that together make the Great Square of Pegasus asterism.

PISCES

ZETA (ζ) PISCIUM 🔭

THE CIRCLET

WIDE VIEW 👁
Pisces is the faintest constellation of the zodiac, narrowly beating Cancer, the Crab. The constellation represents a pair of mythical fishes tied together with ribbon.

PEGASUS (Peg)

WIDTH / DEPTH 🖐🖐🖐 / 🖐

SIZE RANKING 7th

FULLY VISIBLE 90°N–53°S

SKY MAP: PEGASUS

Strangely, one of the four stars that makes up the Square of Pegasus, Alpheratz, is no longer considered to be part of Pegasus; it now belongs to the neighboring constellation of Andromeda.

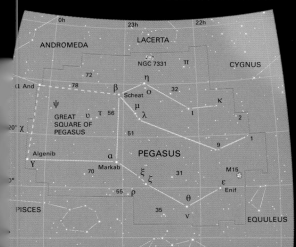

OBSERVING PEGASUS

MAJOR STARS	MAGNITUDE	FEATURES OF INTEREST
Alpha (α) Pegasi *Markab*	2.5	It is believed that this star will soon begin to turn into a red giant, as it is nearing the end of its main life.
Beta (β) Pegasi *Scheat*	2.5	An irregular variable star 95 times the diameter of the Sun.
Pi (π) Pegasi	4.3	A wide double star with an unrelated companion of mag. 5.6 10' away. Their colors are yellow and white.
51 Pegasi	5.5	The first Sun-like star discovered to be orbited by a planet.

NOTABLE OBJECTS	MAGNITUDE	FEATURES OF INTEREST
M15	6.2	A fine globular cluster at a distance of 33,600 light-years from Earth.

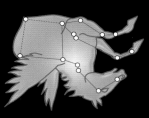

In **ancient Greek** mythology, **Pegasus** is the **winged** steed of the hero **Bellerophon.**

PISCES (Psc)

WIDTH / DEPTH 🖐🖐🖐 / 🖐🖐🖐

SIZE RANKING 14th

FULLY VISIBLE 83°N–56°S

SKY MAP: PISCES

The ecliptic crosses Pisces, so it is visited by planets and the Moon. Pisces also contains the point where the Sun annually crosses the celestial equator into the northern hemisphere.

OBSERVING PISCES

MAJOR STARS	MAGNITUDE	FEATURES OF INTEREST
Alpha (α) Piscium *Alrescha*	4.1	A close pair of stars of 4th- and 5th-magnitudes, 140 light-years away from us.
Beta (β) Piscium *Fum al Samakah*	4.5	A blue star 493 light-years away and found just outside the Circlet.
Zeta (ζ) Piscium	5.3	A double star with a mag. 6.5 companion 25" away. The stars are bluish and white.
Rho (ρ) and 94 Piscium	5.4	A wide naked-eye double star with the mag. 5.6 star 94 Piscium 7' 30" away. The star colors are yellow and golden.

CELESTIAL CIRCLE 👁

The seven stars of the Circlet form the body of the southerly fish, and are of 4th and 5th magnitudes. TX Piscium, here the furthest left of the ring, is a red giant that varies irregularly between magnitude 4.8 and 5.2, appearing noticeably orange in binoculars.

STARHOPPING FROM THE SQUARE OF PEGASUS

Even though the Square of Pegasus is not composed of particularly bright stars, it stands out well in an otherwise very empty part of the night sky.

The Square of Pegasus is formed by three stars of between second and third magnitude—Alpha (α), Beta (β), and Gamma (γ) Pegasi—that lie within Pegasus, and a fourth star, Alpha (α) Andromedae, from the neighboring constellation of Andromeda.

1 TO CYGNUS

Draw an imaginary line across the diagonal of the Square of Pegasus, starting from Algenib, or Gamma (γ) Pegasi—which is located on the winged horse's back—through Beta (β) Pegasi, Scheat, on the opposite side of the Square. Extend this line for twice the distance already traveled and you will reach the star Deneb, the leading star of Cygnus, located on the Swan's tail. Deneb is also one of the stars of the famous Summer Triangle asterism (see pp. 106–107).

CYGNUS

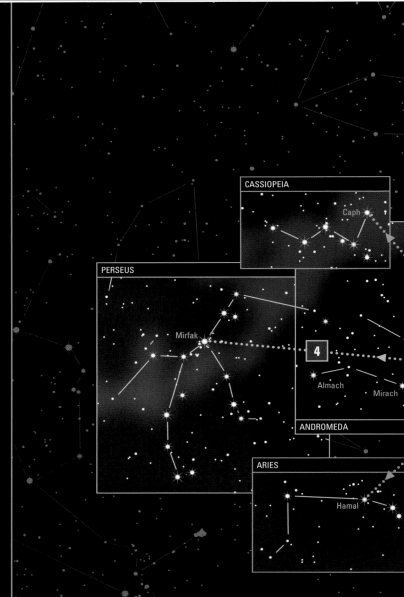

2 TO CASSIOPEIA

As before, begin this starhop from Algenib, which is located in one corner of the Square of Pegasus. Draw a line from here straight up one side of the Square—which is the width of one fist held at arm's length—through Alpheratz, or Alpha (α) Andromedae. Extend the line to the next bright star: this will be Caph, or Beta (β) Cassiopeiae, which lies at the "end" of the distinctive W-shaped constellation of Cassiopeia.

CASSIOPEIA

3 TO PISCIS AUSTRINUS

Begin this journey at Beta (β) Pegasi, or Scheat, the brightest star in Pegasus, which lies in one corner of the Square. Make an imaginary line from here along one edge of the Square to Alpha (α) Pegasi, Markab; extend the line over three times the distance between the first two stars. This will take you to the constellation of Piscis Austrinus, the Southern Fish, with its bright leading star, Fomalhaut.

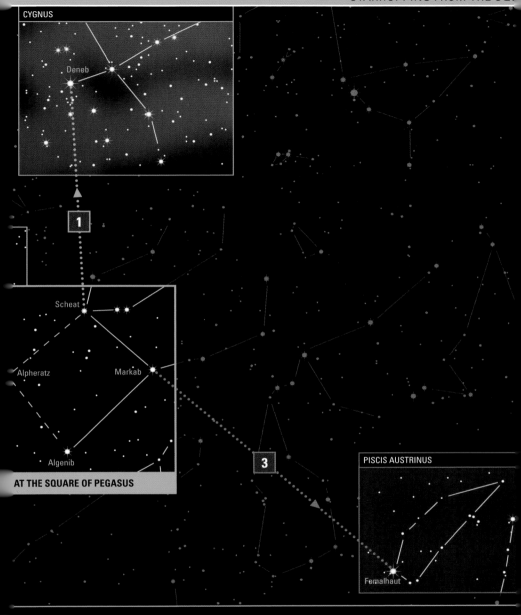

CYGNUS

Deneb

1

Scheat

Alpheratz   Markab

Algenib

AT THE SQUARE OF PEGASUS

3

PISCIS AUSTRINUS

Femalhaut

ANDROMEDA AND PERSEUS

at Markab, Alpha (α) Pegasi, at one
of the Square, journey diagonally across
are through Alpheratz, Alpha (α)
edae, into Andromeda itself. This line will
past two brightish stars: Mirach, Beta (β)
edae (magnitude 2.5), and Almach,
(γ) Andromedae (magnitude 2.2). Further
g this imaginary trail is the bright star

5 TO ARIES

Starting from Beta (β) Pegasi
one corner of the Square of P
imaginary line through Alpher
Andromedae. Extend the line
distance already traveled to r
(α) Arietis, in Aries. Look out
constellation of Triangulum (s
adjacent to Aries, which cont

ANDROMEDA

Andromeda does not really have much of a pattern; her four brightest stars just form an uneven line across the sky. But she has a showpiece: the great Andromeda Galaxy.

Andromeda's head is marked by Alpheratz, Alpha (α) Andromedae, one of the stars that form the Great Square of Pegasus (see p.112)—an easy way to locate the constellation and a great starting point from which to "starhop" to M31, the famous Andromeda Galaxy. Visible to the naked eye, M31 is captivating through binoculars or a telescope, but don't let it distract you from some of the other varied objects in this constellation. Gamma (γ) Andromedae, or Almach, an orange giant with a blue companion, is regarded by many as the second greatest double star in the night sky, after Albireo in Cygnus. The open star cluster NGC 752 is a good target for binoculars. With a telescope the clear blue disc of NGC 7662, the Blue Snowball planetary nebula, is readily visible.

The **Andromeda Galaxy** is the closest major galaxy to the **Milky Way**, and twice as big.

Starhopping within Andromeda

Stargazing works best when it is possible to starhop to locate an object, and from Andromeda it is an easy hop to M31 using this process. Starting from Alpha (α) Andromedae in the Great Square of Pegasus, hop north to Delta (δ), then on to Beta (β) where the track turns in a different direction toward Mu (μ), ending up at Nu (ν). From here the Andromeda Galaxy is merely a small jump. The open star cluster NGC 752 can be found by starhopping directly south from Gamma (γ) Andromedae—Andromeda's left foot—to Deltoton (Beta (β) Trianguli) in the constellation Triangulum. The cluster, which is spread out over quite a wide area of sky, is midway between the two

M31, ANDROMEDA GALAXY 👁

ALPHERATZ

ALMACH ⛢

WIDE VIEW 👁
Andromeda is one of the original Greek constellations. Its three brightest stars represent the princess's head (Alpha), her pelvis (Beta), and her left foot (Gamma).

ANDROMEDA (And)

WIDTH / DEPTH	🖐🖐🖐 / 🖐🖐
SIZE RANKING	**19th**
FULLY VISIBLE	**90°N–37°S**

SKY MAP

Upside-down on star maps, Andromeda is seated, arms outstretched and chained to the rocks as a sacrifice to the sea-monster Cetus, whose own constellation lurks to one side, beyond Pisces, while Perseus flies to her aid from the other.

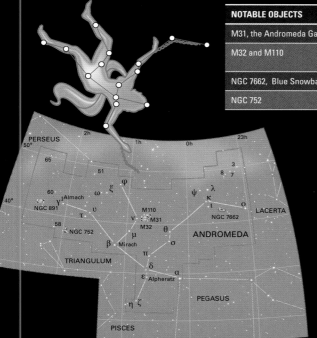

OBSERVING ANDROMEDA

MAJOR STARS	MAGNITUDE	FEATURES
Alpha (α) Andromedae *Alpheratz*	2.1	A bluish star that sits 97 light-years away.
Beta (β) Andromedae *Mirach*	2.1	A red giant wih a diameter 90 times that of the Sun.
Gamma (γ) Andromedae *Almach*	2.3	A fantastic double star with a mag. 5.1 companion sitting 10" away. The stars are golden and green.
56 Andromedae	5.7	A wide double star with a mag. 5.9 companion sitting 3' 20" away. The stars are yellow and orange.

NOTABLE OBJECTS	MAGNITUDE	FEATURES
M31, the Andromeda Galaxy	4.5	A spiral galaxy that neighbors the Milky Way.
M32 and M110	8.1 and 8.9	Dwarf elliptical galaxies that are satellites of M31. They can be seen in a small-to-medium telescope.
NGC 7662, Blue Snowball	9.2	A planetary nebula, good through a telescope.
NGC 752	5.5	An open cluster of around 70 stars in a loose group.

STAR CHAIN 👥

NGC 752 is a loose star cluster that looks great in binoculars, appearing next to a curving chain of brighter, colored stars to its south. A telescope may not do the cluster as much justice as binoculars, as it is the majesty of a wide field that provides the best view.

M31, the Andromeda Galaxy 👁

From a dark location, M31, the Andromeda Galaxy, looks to the naked eye like a slightly oval misty patch. At around 2.5 million light-years away, M31 is one of the furthest objects that can be seen with the naked eye. A nearby galaxy, M33 in Triangulum, is technically further and is supposedly a naked-eye object, yet even seasoned stargazers find its faintness an impossible challenge. The brightness of the Andromeda Galaxy makes it visible in binoculars even from towns and cities. This is not surprising, considering that the most recent estimates suggest the galaxy has over one trillion stars shining out into the Universe.

GALACTIC MIGHT
If our eyes were capable of taking in the full majesty of the Andromeda Galaxy, it would appear six times the width of the full Moon.

GALACTIC NEIGHBOR
This magnificent deep-sky object is the Andromeda Galaxy. It is the closest major galaxy to our own, and the largest member of the Local Group of galaxies, being twice as wide as the Milky Way.

ARIES AND TRIANGULUM

Both Aries, the Ram, and Triangulum, the Triangle, are not prominent constellations, yet their histories date back thousands of years.

In ancient times, Aries was designated the first constellation of the zodiac—the 12 constellations that sit behind the ecliptic, the yearly path taken by the Sun across the sky. Triangulum also has a long history and was originally given the name of Deltotron due to its resemblance to the Greek capital letter Delta (Δ). Aries is an inconspicuous constellation that lies between Pisces and Taurus. Its most recognizable features are three stars that are near its border with Pisces, forming the Ram's "head": Alpha (α), Beta (β), and Gamma (γ) Arietis—Hamal, Sheratan, and Mesarthim. Aries was once the home of the vernal equinox in the age of the ancient Greeks. The constellation of Triangulum is also unspectacular— consisting of only three stars, none of which are first-magnitude—although it is home to M33, a spiral galaxy that is one of the largest in our galaxy's Local Group. For a period, Triangulum was neighbor to a smaller partner, Triangulum Minor, as well as Musca Borealis, the Northern Fly. Neither of these figures are now in use; they are two of several insignificant constellations introduced but later discarded by astro-cartographers.

M33, the Triangulum Galaxy ✦
Almost certainly the farthest object that it is possible to see with the naked eye, M33, the Triangulum Galaxy, lies a staggering distance of 3 million light-years from Earth. When we consider that light takes just over eight minutes to travel from the Sun to the Earth, it puts the distance between us and M33 into some perspective. It is the third-largest member of the Local Group of galaxies, after the Andromeda Galaxy and our own Milky Way. M33 is a close neighbor of the Andromeda Galaxy (see pp.116–117) and is affected by the latter's gravity, potentially even orbiting it. M33 covers about the same area of sky as the full Moon, but because its light is so spread out, a clear, dark night is needed to view it.

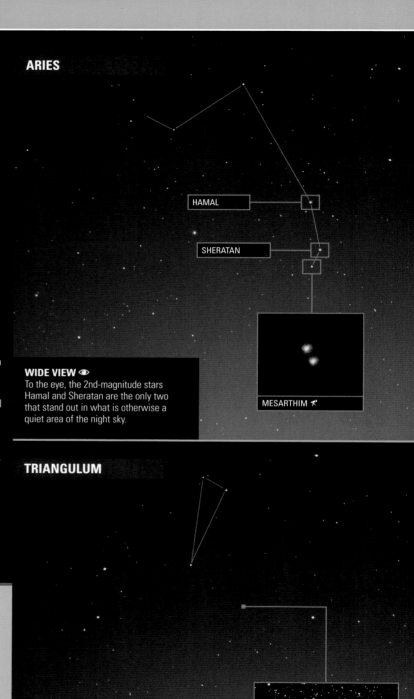

ARIES

HAMAL

SHERATAN

MESARTHIM ✦

WIDE VIEW 👁
To the eye, the 2nd-magnitude stars Hamal and Sheratan are the only two that stand out in what is otherwise a quiet area of the night sky.

TRIANGULUM

M33, TRIANGULUM GALAXY ✦

WIDE VIEW 👁
On clear nights, M33 can be seen through binoculars and telescopes as a large, pale patch the size of the full Moon—larger telescopes are needed to see its spirals.

ARIES (Ari)

WIDTH / DEPTH	
SIZE RANKING	**39th**
FULLY VISIBLE	**90°N–58°S**

SKY MAP: ARIES

The ecliptic runs through Aries. This means that the planets and the Moon appear there occasionally, which helps when it comes to locating the Ram.

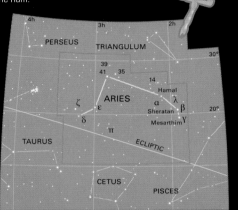

OBSERVING ARIES

MAJOR STARS	MAGNITUDE	FEATURES OF INTEREST
Alpha (α) Arietis *Hamal*	2.0	A large star with a diameter 15 times greater than that of the Sun.
Beta (β) Arietis *Sheratan*	2.7	A bluish star around 60 light-years from Earth.
Gamma (γ) Arietis *Mesarthim*	4.8	A good double star for viewing in a telescope, with its almost identical mag. 4.8, white companion 10" away.

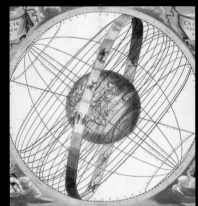

FIRST POINT OF ARIES

Around 2,000 years ago, Aries held a very important position. The vernal equinox—the point where the Sun crosses the celestial equator moving from south to north—lay within Aries. It was called the First Point of Aries and marked the start of spring in the northern hemisphere and fall in the southern. This point is no longer in Aries, but has moved into Pisces, toward Aquarius.

TRIANGULUM (Tri)

WIDTH / DEPTH	
SIZE RANKING	**78th**
FULLY VISIBLE	**90°N–52°S**

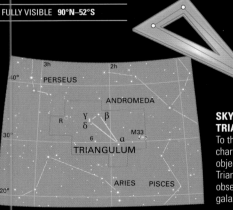

SKY MAP: TRIANGULUM

To the right of the chart is the celestial object that makes Triangulum worth observing: the spiral galaxy M33.

OBSERVING TRIANGULUM

MAJOR STARS	MAGNITUDE	FEATURES OF INTEREST
Alpha (α) Trianguli *Rasalmothallah*	3.4	A yellowish star with a diameter nearly three times larger than the Sun's.
Beta (β) Trianguli *Deltoton*	3.0	About 125 light-years away, and the brightest star in the constellation.
R Trianguli	5.4	A long-period Mira-type variable whose magnitude changes from 5.4 to 12.6 over 267 days.

NOTABLE OBJECTS	MAGNITUDE	FEATURES OF INTEREST
M33, the Triangulum Galaxy	5.7	The oval shape of this spiral galaxy can be seen in binoculars.

SPIRAL IN THE SKY

M33 is an example of a flocculent spiral—a galaxy with arms that divide and separate into patches. It is more typical of a spiral galaxy than the Andromeda Galaxy.

Triangulum was known to the ancient **Greeks**, who visualized it as the **Nile Delta** or the island of **Sicily**.

PERSEUS

Sword held aloft, Perseus, the Hero, is an easily recognizable constellation. Even in this relatively busy part of the sky, it is full of interesting stars and deep-sky objects to observe.

There are a great variety of objects in Perseus. Many are visible to the eye, but this constellation also wonderfully repays a slow sweep across it with binoculars. Sitting over the Milky Way, it contains nebulae and clusters in abundance. The leading star Alpha (α) Persei, Mirfak, is the centerpiece of a collection of hot young stars known as the Perseus Moving Cluster, around 600 light-years away. The M34 open cluster is just visible to the unaided eye from very dark sites, but only binoculars will reveal its 80 or so stars. The Sword Handle double open star cluster is stunning in binoculars. Perseus also produces one of the best annual meteor showers, the Perseids. At their peak around August 12–13, up to 80 shooting stars can be seen every hour.

The twin open **clusters** of the **Sword Handle** form one of the **marvels** of the night.

The Demon's eye 👁

One of the most interesting stars in Perseus is Beta (β) Persei, or Algol, a star that demonstrates that the night sky is not a fixed and unchanging place. Algol is an eclipsing binary, a type of variable star whose brightness can be seen to change over the course of just one evening. Algol consists of two stars orbiting one another over nearly three days, partially along our line of sight. This means that each star "eclipses" the other as they go around. In the classical figure of the constellation, Algol ("El Ghoul") represents the eye of the Gorgon Medusa (see *The legend of Perseus*, opposite), and because of the real star's significant magnitude change, it is nicknamed the Winking Demon.

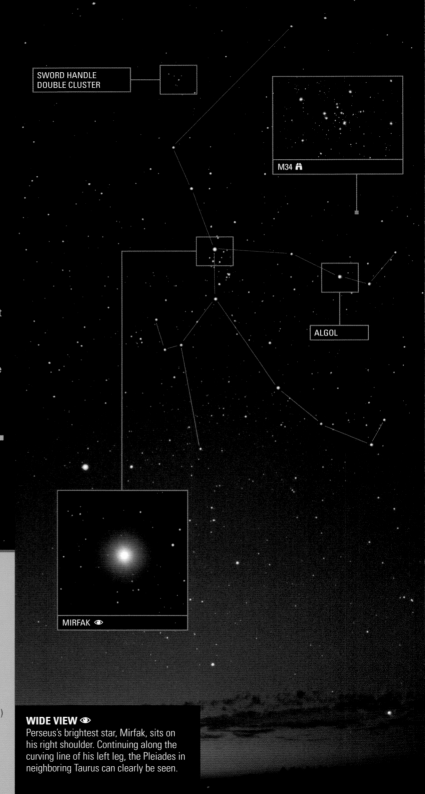

SWORD HANDLE DOUBLE CLUSTER

M34 🏃

ALGOL

MIRFAK 👁

WIDE VIEW 👁
Perseus's brightest star, Mirfak, sits on his right shoulder. Continuing along the curving line of his left leg, the Pleiades in neighboring Taurus can clearly be seen.

PERSEUS (Per)

WIDTH / DEPTH	
SIZE RANKING	24th
FULLY VISIBLE	90°N–31°S

SKY MAP

Perseus sits in a fairly busy part of the sky between the constellations of Cassiopeia, the Queen, and Auriga, the Charioteer. Striding into battle with the sea-monster Cetus, he brandishes his sword in hs right hand, while in his left he holds the decapitated head of Medusa.

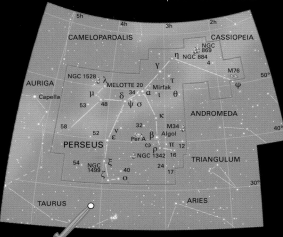

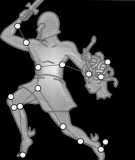

OBSERVING PERSEUS

MAJOR STARS	MAGNITUDE	FEATURES
Alpha (α) Persei *Mirfak*	1.8	A yellow-white star with a diameter 62 times larger than that of the Sun.
Beta (β) Persei *Algol*	2.1	An eclipsing variable star that dims for 10 hours to mag. 3.4 precisely every 2 days, 20 hours, and 4 minutes.
Eta (η) Persei *Miram*	3.8	A double star with a mag. 8.5 companion. The stars are orange and blue.

NOTABLE OBJECTS	MAGNITUDE	FEATURES
M34	5.6	An open star cluster around 1,400 light-years away.
NGC 869 & 884, The Sword Handle	4.3	A magnificent double star cluster; both clusters lie between 6,800 and 7,200 light-years away.
NGC 1528	6.4	An open star cluster formed of around 50 stars; a good target for a small telescope.

THE SWORD HANDLE 👁

Although easy to see with the naked eye as two compact misty patches in the Milky Way, binoculars or a wide-field telescope bring out the beauty of this pair of star clusters, each the size of the full Moon. Seen together, NGC 869 and 884 are an incredible sight.

The legend of Perseus

Several constellations in this part of the night sky are linked in one classical story. Queen Cassiopeia, wife of King Cepheus, was so boastful about her own beauty and that of their daughter Andromeda that it angered the sea god Poseidon. To appease him, Andromeda was chained to a rock as a sacrifice to the sea monster, Cetus. The hero Perseus, meanwhile, was traveling home after slaying the Gorgon Medusa—from a drop of whose blood had sprung the winged horse Pegasus. Spying Andromeda in peril, Perseus rescued her by turning Cetus to stone—the fate of any being who looked into the eyes of Medusa, whose severed head Perseus carried in a bag.

HEROIC RESCUE

Andromeda married Perseus, her rescuer, and on her death was placed in the sky near him and her mother, Cassiopeia, by the goddess Athena.

AURIGA

Auriga, the Charioteer, forms a bright, simple shape that is easily located in the evening winter sky by observers in the northern hemisphere.

There are constellations whose shape is readily reflected in their name, such as Crux and Leo, and then there are those that require a lot more imagination. Auriga most certainly falls into this latter category—as you would be hard-pressed to see its main stars as a classical charioteer—but at least its component stars are bright and easily observable. They are led by the marvelous Alpha (α) Aurigae, or Capella, the sixth-brightest star in the entire night sky. However, as with so many stars, appearances can be deceptive; this is not a single star but four. The light from Capella is actually the combined light from two bright, yellowish stars that orbit around one another every 104 days. These are big stars—each of their diameters is around 10 times larger than the Sun's—and both are in turn orbited by a red dwarf star. Capella is a Latin name that means "she-goat," and there are three smaller stars to its south—Epsilon (ε), Zeta (ζ), and Eta (η) Aurigae— known as the Kids (of the Goat). As Auriga lies over the Milky Way, there are deep-sky objects to be seen with binoculars or a telescope. It boasts a chain of star clusters—M36, M37, and M38— and outside the main constellation lies NGC 2281, a loose cluster of several groups of stars.

M36, M37, and M38 📷

These open star clusters naturally lend themselves to being observed and discussed together, as they are closely grouped. They sit in a crooked line that runs across Auriga, perpendicular to an imaginary line from Theta (θ) Aurigae to Beta (β) Tauri in Taurus. M36 is the smallest of the three, with a rather scattered appearance. The largest, at two-thirds the apparent size of the full Moon, M37 is also the brightest, simply because it contains more stars than the others. It has been suggested that the stars of the last cluster, M38, roughly resemble the Greek letter pi (π). Wide-angle binoculars are ideal for viewing these clusters, as all three can just be seen together in the same field of vision as misty patches.

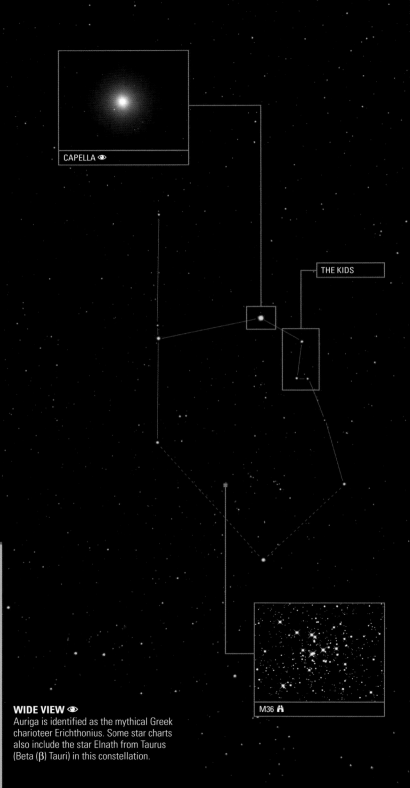

CAPELLA 👁

THE KIDS

M36 📷

WIDE VIEW 👁
Auriga is identified as the mythical Greek charioteer Erichthonius. Some star charts also include the star Elnath from Taurus (Beta (β) Tauri) in this constellation.

AURIGA (Aur)

WIDTH / DEPTH	·· / ··
SIZE RANKING	**21st**
FULLY VISIBLE	**90°N–34°S**

SKY MAP

Auriga is unusual among constellations in that it is necessary to use a star from a neighboring constellation, Beta (β) Tauri in Taurus, to complete the classical figure. Even then, the interpretation of this shape as "the Charioteer" requires something of an imaginative leap.

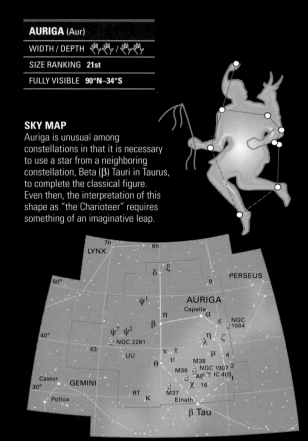

FIERY NEBULA

The star-forming nebula IC 405, or the Flaming Star Nebula, is lit up by the 6th-magnitude star AE Aurigae. This is a "runaway" star that may have been ejected during a collision of two binary star groups. The collision is also credited with ejecting Mu (μ) Columbae.

OBSERVING AURIGA

MAJOR STARS	MAGNITUDE	FEATURES
Alpha (α) Aurigae *Capella*	0.1	A 42-light-year distant system of four stars.
Beta (β) Aurigae *Menkalinan*	1.9	The light of this star is made of two bright stars that orbit each other every four days.
Epsilon (ε) Aurigae *Al Anz*	3.1	An eclipsing binary that dips to mag. 3.8 every 27 years.
RT Aurigae	5.0	A Cepheid variable that fades to mag. 5.8 every 3½ days.

NOTABLE OBJECTS	MAGNITUDE	FEATURES
M36	6.0	This open star cluster, together with M37 and M38 (see below), can be seen in a singular binocular field.
M37	5.6	A large open star cluster of around 500 stars.
M38	6.4	An open star cluster 4,200 light-years from Earth.
NGC 2281	5.4	A loose open star cluster of around 30 stars.

Epsilon (ε) Aurigae 👁

Epsilon (ε) Aurigae, or Al Anz, is an ordinary-looking star that is one of the "Kids" of Capella. As is often the case with astronomical discoveries, it was not until the invention of the telescope, and developing observation techniques, that a truer picture of its activities came to light. Al Anz is a fine example of this ongoing process, for while we have uncovered much about this star, there is still a lot to learn. We now know it is a binary system: a bright supergiant orbited by a mysterious "dark" partner—possibly shrouded by a dusty disc—that eclipses it every 27 years, the longest interval of any known eclipsing binary. Further study should reveal more about this "dark star."

ECLIPSING BINARY STARS

Epsilon (ε) Aurigae is a supergiant orbited by a smaller "dark star," possibly surrounded by a dust cloud, as shown (bottom) in this illustration.

TAURUS

Taurus, although not looking much like a Bull, does contain some really interesting sights, including the splendid star group of the Pleiades, or Seven Sisters, and the Crab Nebula.

Taurus is one of the oldest constellations, known to the ancient Egyptians and the Babylonians and possibly to prehistoric peoples. Fittingly, it contains a magnificent dying star, the Crab Nebula. Its leading star, Aldebaran, is also aging, although this orange-red giant, with a diameter around 45 times that of the Sun, is still the 13th-brightest star in the sky. The eye of the Bull, Aldebaran is also the centerpiece of a star group known as the Hyades, but it is another star cluster, the Pleiades, that is the real glory of Taurus. One of the most instantly recognizable objects in the night sky, through binoculars, this glittering smudge resolves itself into a packed star cluster on a milky background, dominated by the bright Seven Sisters of Greek mythology.

Prehistoric paintings in France's **Lascaux Caves** show a **bull** surrounded by **stars**.

M45, the Pleiades (Seven Sisters) 👁
The Pleiades is one of the jewels of the night sky, appearing to the naked eye as a small, sparkling group of stars. These are actually just the brightest members of a relatively close open cluster of possibly 500 stars. This group is known as the Seven Sisters (although it contains nine named stars), but differing numbers of stars may be visible to the naked eye on any given night. Light pollution may lower the number to four or five, while viewing in a dark, out-of-town location could increase the number to more than 10. With binoculars, the fainter members come into view and the star count increases to dozens, bringing out the full glory of the Pleiades.

M45, THE PLEIADES 👁

THETA-1 (θ¹) TAURI 👁

WIDE VIEW 👁
This large and prominent constellation is easily traced from the Pleiades through the V-shaped Hyades, up to the Bull's two star-tipped horns.

TAURUS (Tau)

WIDTH / DEPTH	🖐🖐🖐 / 🖐🖐
SIZE RANKING	**17th**
FULLY VISIBLE	**88°N–58°S**

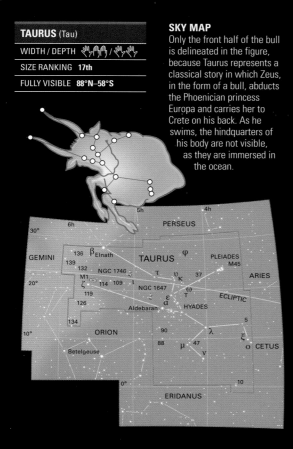

SKY MAP

Only the front half of the bull is delineated in the figure, because Taurus represents a classical story in which Zeus, in the form of a bull, abducts the Phoenician princess Europa and carries her to Crete on his back. As he swims, the hindquarters of his body are not visible, as they are immersed in the ocean.

OBSERVING TAURUS

MAJOR STARS	MAGNITUDE	FEATURES
Alpha (α) Tauri *Aldebaran*	0.9	This large orange star forms the eye of the Bull in the constellation's classical design.
Beta (β) Tauri *Elnath*	1.7	A blue star about 131 light-years away.
Zeta (ζ) Tauri	3.0	A variable star dimming irregularly from mag. 2.9 to 3.2.
Theta-1 Tauri (θ¹)	3.8	A double star with a mag. 3.4 companion, Theta-2 (θ²). They appear yellow and white.
Sigma-1 Tauri (ς¹)	5.1	A double star with a mag. 4.7 companion, Sigma-2 (ς²). Both stars are white.

NOTABLE OBJECTS	MAGNITUDE	FEATURES
M1, the Crab Nebula	8.4	A supernova remnant at a distance of 6,300 light-years.
M45, the Pleiades	1.5	An open star cluster best viewed with binoculars.

THE FOLLOWER 👁

Aldebaran can be seen as the orange "eye" within the V-shaped star group that forms the bull's face, the Hyades, although it is not physically associated with the group, lying less than half their distance from Earth. Aldebaran translates from Arabic as "the Follower," because it seems to follow the Pleiades (top right) around the sky.

M1, the Crab Nebula 🔭

The first object in Charles Messier's famous deep-sky catalog (see p.42) lies in Taurus: M1, the Crab Nebula. It is the remnants of a star that exploded in July 1054. Chinese astronomers of the time recorded that the blast was visible as an incredibly bright star in the daytime skies for 23 days. Since then, the debris of this supernova has been, and still is, expanding through space, so our present view of the star's remains is only temporary. The Crab's apparent magnitude is 8.4, but it is a very small and faint object until viewed in a larger telescope. Locating M1 is easy: it sits just above the star Zeta (ζ) Tauri that marks the tip of the Bull's lower horn.

BEAUTIFUL IN DEATH

M1 is made up of gas filaments radiating from a 1,000-year-old explosion. The original star has become a tiny but incredibly dense, fast-spinning pulsar.

GEMINI

Gemini, the Twins, forms part of a bright set of constellations that makes the night sky so fascinating for skywatchers in the northern hemisphere over the cold winter months.

Gemini is quite an easy constellation to identify with the eye, as its stars form a simple pattern of two parallel lines, led by the two main stars that give the constellation its name. Alpha (α) and Beta (β) Geminorum, better known as Castor and Pollux, appear close and almost twin-like in the sky. As is sometimes the case the leading, or alpha star—in this case Castor—is strangely fainter than its brother (see below). Castor has a blue-white color, and so makes a good contrast with its pale-orange twin. Using a telescope will reveal Gemini to be a constellation scattered with interesting objects, including double stars and a fine open star cluster, M35. On really clear, dark nights, M35 can just be seen with the naked eye, but binoculars will make it come alive.

The ancient **Greeks** believed that **Zeus** turned the **twins** Castor and Pollux into stars.

Castor 👁
Although they are referred to as twins, the stars Castor and Pollux are far from identical. Castor is a truly remarkable star, or rather system of stars. The single point of light that is visible to the naked eye becomes two bright blue-white points of apparent magnitudes 1.9 and 2.9 when seen through a small telescope. These two stars are also gravitationally bound, which means they orbit each other— approximately once every 468 years. However, that is not the end of the story, for each of these stars is also a double. In addition, a nearby 9th-magnitude red dwarf star has also been found to be part of this group, and it too is a double star. This brings the total number of stars in the Castor system to six.

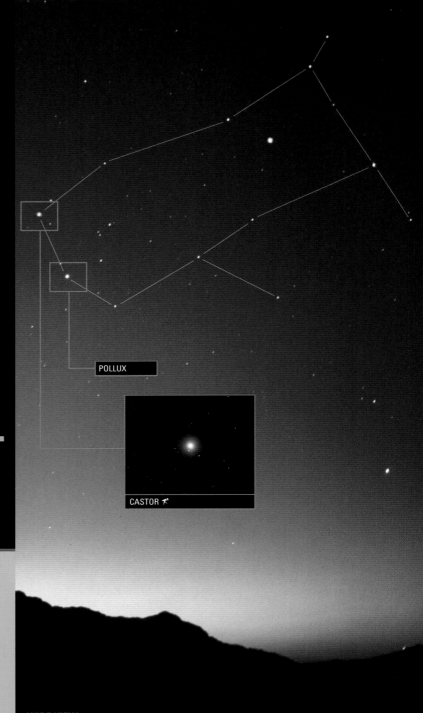

POLLUX

CASTOR 🏃

WIDE VIEW 👁
The "twins" Castor and Pollux lie side-by-side between Taurus and Cancer. The bright "star" here in the center of the constellation is actually the planet Saturn.

GEMINI (Gem)

WIDTH / DEPTH	
SIZE RANKING	30th
FULLY VISIBLE	90°N–55°S

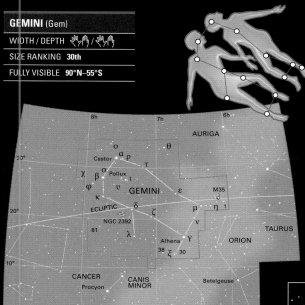

OBSERVING GEMINI

MAJOR STARS	MAGNITUDE	FEATURES
Alpha (α) Geminorum *Castor*	1.6	A bright star around 50 light-years away.
Beta (β) Geminorum *Pollux*	1.15	A large star whose diameter is eight times greater than that of the Sun.
Zeta (ζ) Geminorum *Mekbuda*	3.6–4.2	A Cepheid variable star (see p.90) with a period of 10.2 days.
Eta (ε) Geminorum *Propus*	3.2–4.2	A semi-regular variable star with a period of about 233 days.

NOTABLE OBJECTS	MAGNITUDE	FEATURES
M35	5.3	An open star cluster that covers an area of sky the size of the full Moon.
NGC 2392, the Eskimo Nebula	8.6	A planetary nebula observable through a telescope as a blue-green oval.

SKY MAP

The twins stand side-by-side across the ecliptic, the sky path of the Sun through the year, which thus places them among the zodiacal constellations. The bright stars Castor and Pollux mark the heads of the brothers, while their feet are bathed in the Milky Way.

TWIN STAR 👁

Beta (β) Geminorum, Pollux, is noticeably warmer in tone than its "twin," Castor, and is significantly closer to the Earth, at only 34 light-years' distance, as opposed to 52.

SCATTERED CLUSTER 👁

Sitting just to the north of Propus, Eta (η) Geminorum, M35 is a great, loose-looking open star cluster. It lies 2,800 light-years from the Solar System with an apparent magnitude of 5.3. Even though this brightness makes it visible to the naked eye, you need binoculars or a telescope to see its 200 or so stars and elongated appearance.

NGC 2392, the Eskimo Nebula 🔭

This planetary nebula is so-called because it has the appearance of a face surrounded by a furry hood. The central bright point is the cause of the nebula—a star very much like the Sun ejected its atmosphere about 10,000 years ago, producing this wondrous structure. The outer envelope shows orange filaments produced by later blasts from the star, which hit knots of denser material flowing away from it, creating an effect similar to the wake from stones thrown in a fast-moving river. The whitish globe at its center is formed of two "lobes" extending from the star's equator, which from our viewpoint, overlap one another.

CELESTIAL PARKA

The Eskimo Nebula, NGC 2392, as imaged by the Hubble Space Telescope, is an amazing planetary nebula that sits 3,800 light-years away from us.

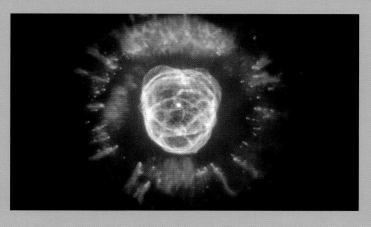

LEO AND CANCER

Here are two constellations sitting side by side that could not be more different. Cancer is built of faint stars, making it rather inconspicuous, while Leo's bright stars dominate their part of the sky.

Leo, the Lion, is one of the few constellations whose star pattern really resembles its classical figure, making it one of the easiest to recognize. It represents the mythological lion that was slain by the Greek hero Hercules in the first of his 12 labors. Surrounded by stars that are not so bright, the shape of this large constellation is easy to pick out in the night sky—especially as its head also contains six stars that form an asterism known as the Sickle. Leo's leading star, Alpha (α) Leonis, or Regulus—also known as Cor Leonis or the Heart of the Lion—sits close to the ecliptic and is one of the few stars that can be covered, or occulted, by the Moon. Cancer, the Crab, is a dim constellation that is easily overlooked, but it is worthy of a mention because it contains one of the finest open star clusters to be found in the night sky. Cataloged as M44, it is named the Beehive, or Praesepe. This fine deep-sky object is one of the few scattered throughout the heavens that can be seen with the naked eye, and this alone means Cancer should not be overlooked when planning an observing session.

M44, the Beehive 👁
Due to its brightness, the Beehive open cluster, or M44, in Cancer has been known for many thousands of years; the Greeks could see it with the naked eye as a fuzzy spot in this faint constellation. Located between Gamma (γ) and Delta (δ) Cancri, it sits approximately 577 light-years from Earth, making it one of our nearest star clusters, and it contains several hundred stars, many of which are doubles. Its alternative name, Praesepe, translates from Latin as "Manger," reflecting the fact that both the Greeks and Romans saw M44 as the manger from which two donkeys ate. The stars just above and below the Beehive—Assellus Borealis and Assellus Australis—are named the Northern and Southern Donkey.

LEO

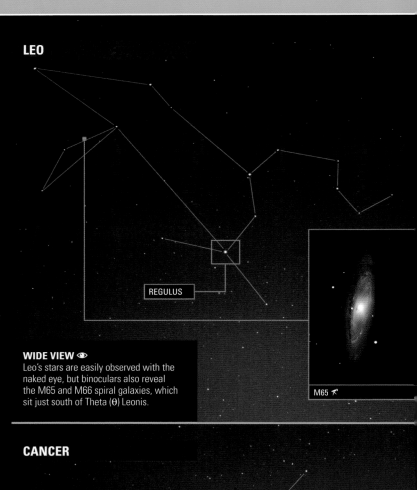

REGULUS

M65 ✈

WIDE VIEW 👁
Leo's stars are easily observed with the naked eye, but binoculars also reveal the M65 and M66 spiral galaxies, which sit just south of Theta (θ) Leonis.

CANCER

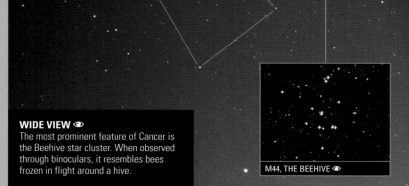

M44, THE BEEHIVE 👁

WIDE VIEW 👁
The most prominent feature of Cancer is the Beehive star cluster. When observed through binoculars, it resembles bees frozen in flight around a hive.

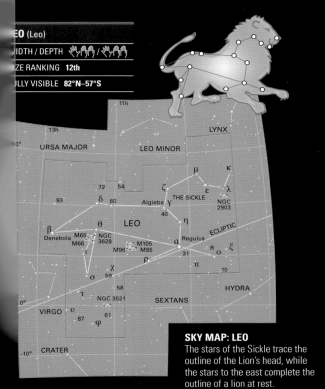

EO (Leo)

DTH / DEPTH	🖐🖐 / 🖐🖐🖐
ZE RANKING	12th
LLY VISIBLE	82°N–57°S

SKY MAP: LEO
The stars of the Sickle trace the outline of the Lion's head, while the stars to the east complete the outline of a lion at rest.

OBSERVING LEO

MAJOR STARS	MAGNITUDE	FEATURES OF INTEREST
Alpha (α) Leonis *Regulus*	1.4	The brightest star in the constellation is white, and sits 78 light-years from Earth.
Beta (β) Leonis *Denebola*	2.2	A bluish star that marks the position of the Lion's tail.
Gamma (γ) Leonis *Algieba*	2.3	A great double star in a telescope with a mag. 3.6 companion 5" away; both are yellow in color.

NOTABLE OBJECTS	MAGNITUDE	FEATURES OF INTEREST
M65	10.3	A spiral galaxy that can be observed with binoculars, but a telescope is better.
M66	8.9	A spiral galaxy that can be found close to M65.

THE LEONIDS 👁

The Leonid meteor shower, which annually peaks around November 17, is so-called because the glowing trails of the meteors radiate out from the constellation of Leo. It is eagerly anticipated every year because it can produce magnificent meteor numbers.

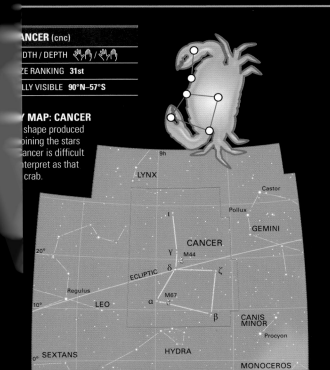

ANCER (cnc)

DTH / DEPTH	🖐🖐 / 🖐🖐
ZE RANKING	31st
LLY VISIBLE	90°N–57°S

Y MAP: CANCER
shape produced
oining the stars
ancer is difficult
terpret as that
crab.

OBSERVING CANCER

MAJOR STARS	MAGNITUDE	FEATURES OF INTEREST
Alpha (α) Cancri *Acubens*	4.3	One of the claws of the crab.
Beta (β) Cancri *Altarf*	3.5	A giant orangey-red star; the brightest in the constellation.
Zeta (ζ) Cancri *Tegmine*	4.7	A double star with a mag. 6.2 companion 5" away.
Iota (ι) Cancri	4.0	A yellow and blue double star with a mag. 6.0 companion 30" away.

NOTABLE OBJECTS	MAGNITUDE	FEATURES OF INTEREST
M44, the Beehive or Praesepe	3.7	A large open star cluster; viewable with the naked eye in good conditions.
M67	6.1	An open star cluster that can easily be found in binoculars.

The **superb** Beehive star cluster, at the **heart** of Cancer, is one of the **classic** amateur targets

MONOCEROS AND CANIS MINOR

With a little imagination, the stars of Monoceros can be visualized as a Unicorn, but with only two significant stars, it is a little more difficult to see Canis Minor as a small dog.

Although the sky is full of constellations, those with bright, recognizable patterns tend to be the most well-known, while the fainter ones can be overlooked. This is the case for Monoceros, the Unicorn. It is lost in the glare of nearby Orion, Canis Major, and Canis Minor, all three of which have brilliant stars that light up the northern hemisphere's winter skies. Delve into Monoceros with a telescope, however, and an amazing collection of objects can be viewed, helped by the fact that the Milky Way runs through the constellation. Canis Minor, meanwhile, is a fairly bare constellation, with nothing more to offer the casual stargazer than its leading star Alpha (α) Canis Minoris, or Procyon. However, this is no ordinary star; it is the seventh-brightest in the entire night sky, and at just over 11 light-years from the Solar System it is also one of the closest stars to us. Procyon forms a large triangle of bright stars with Betelgeuse (in Orion) and Sirius (in Canis Major). This is useful for locating Monoceros, which lies in the center of this triangle.

The Rosette Nebula and NGC 2244

NGC 2244 is a sparkling open cluster consisting of an elongated group of stars of 6th magnitude and fainter. This cluster was formed about four million years ago from the gas and matter of the glorious Rosette Nebula that surrounds it. Through binoculars, the stars appear in a parallelogram and, if conditions are perfect, the Rosette Nebula—which resembles a rose in CCD images—may itself be seen as a faint, ghostly glow around them. It appears around twice the apparent size of the full Moon, and its red glow can be captured with photography. The whole nebula is approximately 130 light-years wide and sits around 5,000 light-years away from the Solar System.

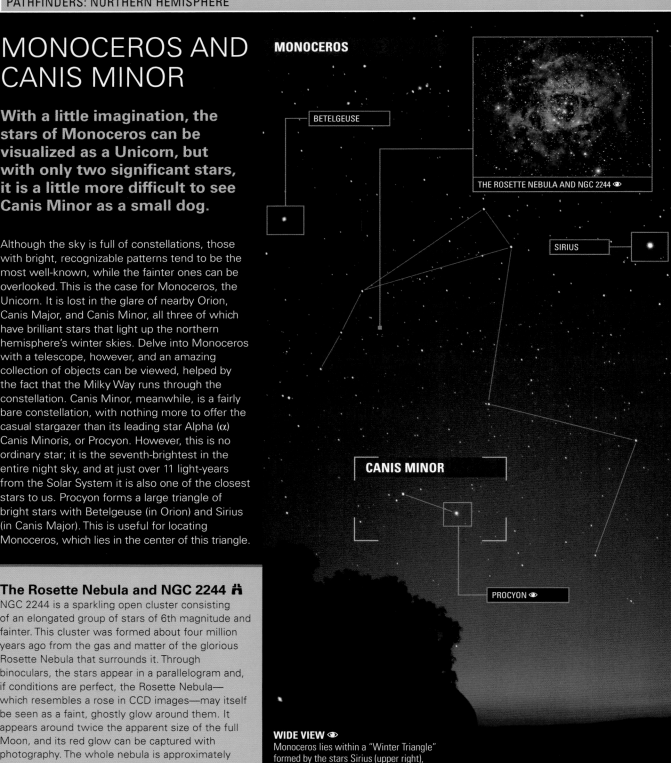

MONOCEROS

BETELGEUSE

THE ROSETTE NEBULA AND NGC 2244 ⊙

SIRIUS

CANIS MINOR

PROCYON ⊙

WIDE VIEW ⊙
Monoceros lies within a "Winter Triangle" formed by the stars Sirius (upper right), Betelgeuse (upper left), and Procyon (Alpha (α) Canis Minoris) in Canis Minor.

MONOCEROS (Mon)

WIDTH / DEPTH	
SIZE RANKING	**35th**
FULLY VISIBLE	**78°N–78°S**

OBSERVING MONOCEROS

MAJOR STARS	MAGNITUDE	FEATURES OF INTEREST
Alpha (α) Monocerotis *Unicorni*	3.9	A yellow star around 144 light-years away.
Beta (β) Monocerotis *Eite*	3.7	An outstanding triple star, the brightest in the constellation. A good target for observers.

NOTABLE OBJECTS	MAGNITUDE	FEATURES OF INTEREST
M50	6.1	An open star cluster that is an easy target for binoculars.
NGC 2237, the Rosette Nebula	9.0	This nebula can just be viewed in good binoculars.
NGC 2244	4.8	An open star cluster that is embedded in the Rosette Nebula (see above).
NGC 2264, the Christmas Tree Cluster	3.9	A group of bright stars over the dark Cone Nebula.

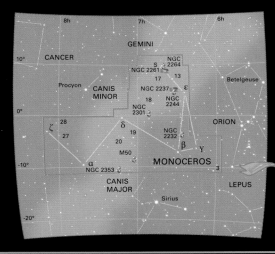

SKY MAP: MONOCEROS

A string of deep-sky objects flows up the chart through this constellation. This is because they are all embedded within the path of the Milky Way.

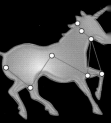

CHRISTMAS LIGHTS

The delightful Christmas Tree Cluster or NGC 2264 (rotated 180° in this image) is an open cluster of around 250 stars, the brightest of which, here at the "trunk," is 15 Monocerotis. Appearing to "point" to the top of the tree is the shadowy form of the Cone Nebula, an immense cloud of dust and gas.

CANIS MINOR (CMi)

WIDTH / DEPTH	
SIZE RANKING	**71st**
FULLY VISIBLE	**89°N–77°S**

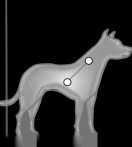

SKY MAP: CANIS MINOR

Even though the Milky Way flows through the south of the constellation, there is nothing of real deep-sky interest to observe here.

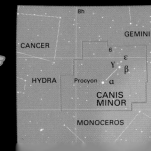

OBSERVING CANIS MINOR

MAJOR STARS	MAGNITUDE	FEATURES OF INTEREST
Alpha (α) Canis Minoris *Procyon*	0.4	A whitish-yellow star sitting 11.6 light-years away from Earth.
Beta (β) Canis Minoris *Gomeisa*	2.9	This star has a diameter four times that of the Sun. Its name means "bleary-eyed woman."

Canis Minor is the smaller of **Orion's** two **dogs**, and consists of little more than its brightest star, **Procyon**

ORION

Orion is one of the finest constellations in the heavens. Its main stars are not only bright, but numerous, and arranged in a wonderfully easy-to-recognize pattern.

Many stargazers get to know this unmissable constellation from the three bright stars known as Orion's Belt. This asterism is very easy to spot because the stars are almost perfectly aligned, equally spaced, and of virtually the same brightness. Just a short way south of the central star of the Belt is one of the gems of the night sky—the Orion Nebula, M42—which forms part of the sword of Orion and is part of a vast cloud of dust and gas inside which stars are being made. M42 can be viewed with the naked eye in a dark location, and looks as if someone has gently smudged the sky. Orion also contains many other interesting objects for all levels of stargazing, and is very useful as a starting point and aid to starhopping around the night sky (see pp.136–137).

This **famous** constellation dominates **northern skies** in the **winter** months.

Betelgeuse 👁

The common name of the red supergiant star Alpha (α) Orionis, Betelgeuse (pronounced "Bet-el-jers" or more commonly "Beetle-juice"), has Arabic origins. The exact translation has been lost in time, but variations include the Central One, the House of Orion, and the Armpit of the Giant. Betelgeuse is one of only a handful of stars whose size has been measured. Most stars are so distant that even the most powerful telescopes reveal no more than a point of light. However, Betelgeuse is enormous; its diameter has been measured at up to 1,000 times that of the Sun. If Betelgeuse were placed in the position of our star, its surface would stretch out to beyond the orbit of Mars.

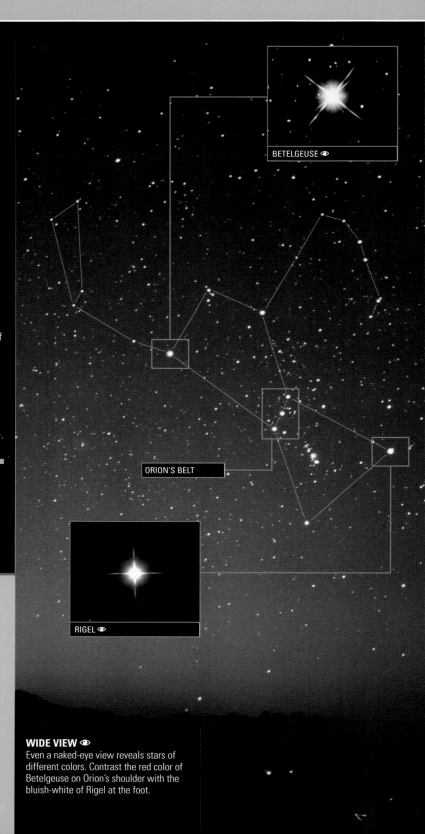

BETELGEUSE 👁

ORION'S BELT

RIGEL 👁

WIDE VIEW 👁
Even a naked-eye view reveals stars of different colors. Contrast the red color of Betelgeuse on Orion's shoulder with the bluish-white of Rigel at the foot.

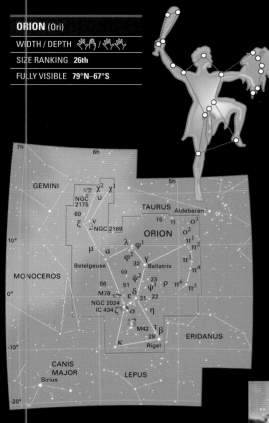

ORION (Ori)

WIDTH / DEPTH	
SIZE RANKING	26th
FULLY VISIBLE	79°N–67°S

OBSERVING ORION

MAJOR STARS	MAGNITUDE	FEATURES
Alpha (α) Orionis *Betelgeuse*	0.0–1.3	A red supergiant, the magnitude of which varies on a six-year cycle.
Beta (β) Orionis *Rigel*	0.2	A blue supergiant around one-tenth the size of Betelgeuse.
Gamma (γ) Orionis *Bellatrix*	1.6	Found on the opposite shoulder of Orion from Betelgeuse.
Delta (δ) Orionis *Mintaka*	2.3	A double star, with a mag. 6.7 companion, at the end of Orion's Belt; appears white and blue.
Epsilon (ε) Orionis *Alnilam*	1.7	The middle star of Orion's Belt.
Zeta (ζ) Orionis *Alnitak*	1.7	A double star at the southeast of Orion's Belt.

NOTABLE OBJECTS	MAGNITUDE	FEATURES
M42, the Orion Nebula	4.0	A big, bright emission nebula, appearing as a smudge of light to the naked eye.
M78	8.3	Reflection nebula: can just be seen using binoculars.
NGC 2169	5.9	An open star cluster of around 30 stars.
B33, the Horsehead Nebula	—	A dark nebula sitting in front of a bright nebula (IC 434), which is easier to locate in the night sky.

SKY MAP

The stars of Orion are interpreted as a hunter, with a club in one hand, and a shield—which he is holding against Taurus, the Bull—in the other. A distinctive line of three stars forms Orion's Belt; hanging from his belt is his sword, which is made up of an area of star clusters and nebulae, including the spectacular Orion Nebula.

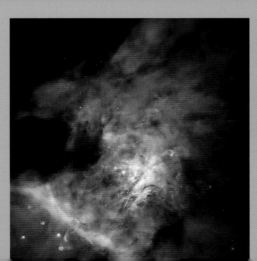

HORSEHEAD NEBULA

This dark nebula, made visible with advanced imaging techniques, is located just below Alnitak, Zeta (ζ) Orionis, in Orion's Belt. The horsehead itself is an extremely dense dust cloud, projecting in front of the pink glow of the bright nebula IC 434.

M42, the Orion Nebula

Even a small telescope will reveal the astonishing beauty of M42, the Orion Nebula. No colors will be seen, but a brighter central area will become apparent; this is the location of four stars known as the Trapezium, due to their visible pattern. Away from them sweeps a delicate curve of mist—a gentle scene that belies what is a turbulent, fiery realm, where vast supersonic "bullets" of gas punch through clouds of hydrogen. At 1,350 light-years from Earth, this distant cloud is believed to be the nursery for more than 2,000 stars and is just the small central part of a vast nebula that covers the entire constellation.

THE ORION NEBULA

M42 is separated by a "lane" of dust (above) from another smaller nebula, M43. A Hubble Space Telescope image shows the Trapezium (left).

STARHOPPING FROM ORION

Orion is a bright and easily recognizable constellation with many geometric configurations, making it a great signpost for finding nearby bright stars and their constellations.

Orion, the Hunter, is depicted as a man holding a club in one hand and a shield in the other. At his shoulder is the red supergiant Betelgeuse, and one of his feet is marked by the luminous-blue supergiant Rigel, which is usually the brightest star in the constellation.

1 TO CANIS MAJOR

Running through the center of Orion is the famous "belt" made up of three stars of almost identical brightness. Draw an imaginary line through the belt, and extend it out from the side

of Orion on which Betelgeuse, the red supergiant, lies. You will then arrive at Alpha (α) Canis Majoris, Sirius, the brightest star in the entire night sky and the leading star of the Great Dog.

CANIS MAJOR

2 TO TAURUS

Once again, the starting point of this starhop is the central Belt of Orion. This time extend an imaginary line in the opposite direction, to the figure's left. The destination is the bright red star Alpha (α) Tauri, Aldebaran, which appears as the "eye" of Taurus, the Bull. Continuing the line roughly onwards, still within Taurus, you will reach the wonderful small open star cluster of the Pleiades (also known as the Seven Sisters)

TAURUS

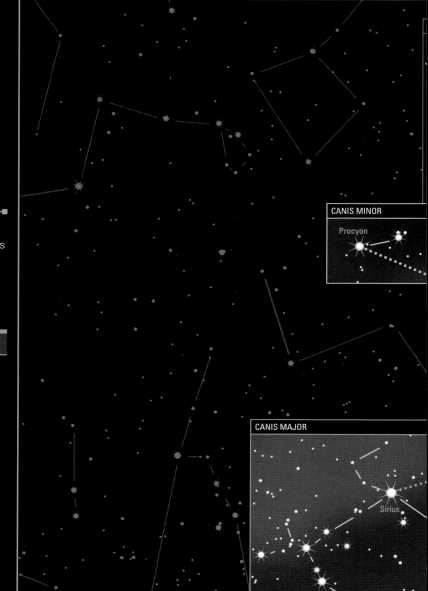

CANIS MINOR
Procyon

CANIS MAJOR
Sirius

3 TO GEMINI

The star that represents the brighter of Orion's two "feet" is the brilliant bluish star Rigel, Beta (β) Orionis. Journey in a roughly straight line from here to Betelgeuse; continue beyond the constellation for approximately twice the distance between Rigel and Betelgeuse and you will arrive at Castor, Alpha (α) Geminorum, and its nearby and slightly brighter twin, Pollux—the two leading stars in Gemini, the Twins.

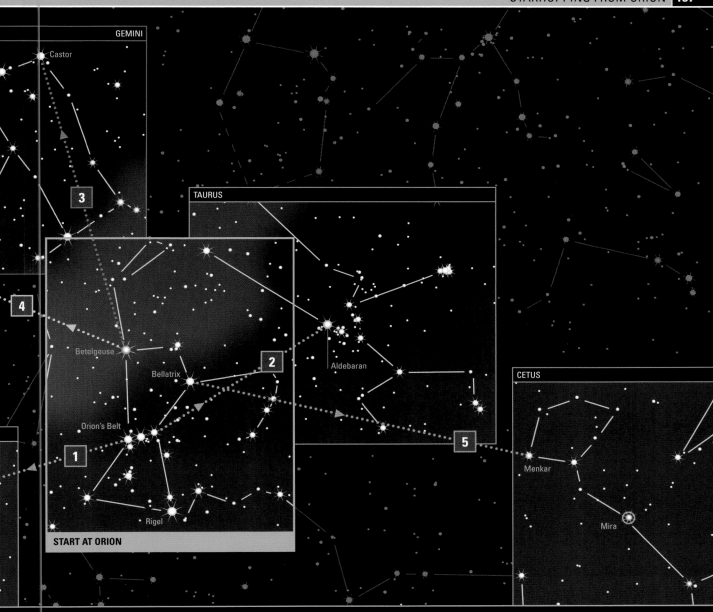

GEMINI

Castor

3

TAURUS

4

Betelgeuse

Bellatrix

2

Aldebaran

CETUS

Orion's Belt

1

5

Menkar

Rigel

Mira

START AT ORION

4 TO CANIS MINOR

At the opposite "shoulder" from Betelgeuse on the figure of Orion is the star Bellatrix, or Gamma (γ) Orionis. With this as the starting point it is very easy to find Procyon, the seventh-brightest star in the night sky. From Bellatrix, draw an imaginary line through the red star Betelgeuse on the other shoulder. With only a slight curve this line extends beyond Orion to Procyon, which is the leading star in Canis Minor, the Little Dog.

CANIS MINOR

5 TO CETUS

Using the same two stars as in the starhop to Canis Minor, it is possible to locate Cetus, the Whale, a constellation usually lost in the glare of nearby stars. This time start from Betelgeuse, pass through Bellatrix and move beyond the curving arc of stars that form Orion's shield. Soon a lone star—and not a particularly bright one at that—will be seen in the emptiness. This is Menkar, Alpha (α) Ceti, at the head of Cetus.

CRUX

Crux, the Southern Cross, does for the southern hemisphere what the Big Dipper does for the northern sky: its shape acts as a signpost to or toward neighboring constellations.

Crux is the smallest of all the 88 constellations, just a closed hand's width across in size, but it nonetheless incorporates one of the finest open star clusters, the Jewel Box, as well as one of the darkest large nebulae, the Coalsack. Given that its cross shape is also one of the most recognizable in the sky, Crux is certainly not a constellation to pass by. Like its northern counterpart, the Big Dipper, Crux can be used as a "signpost"; its "pointers"—the stars Gamma (γ) Crucis and Alpha (α) Crucis, on the longer axis of the cross—can be used to find the south celestial pole, around which the southern sky revolves.

Crux is easily recognizable as its pattern is displayed on the flags of Australia, New Zealand, Samoa, and Papua New Guinea.

The Coalsack 👁

This smudgy cloud of interstellar dust looks almost like a hole that has been roughly cut out of the richly flowing Milky Way. However, the Coalsack is not an absence of stars, but rather is formed by a thick cloud of dust and gas that blocks out the starlight from behind. This is an example of a dark nebula, and quite a close one too, at around 600 light-years away from the Solar System. Dark nebulae are essentially exactly the same as bright nebulae, such as the famous M42 in Orion, except that there are no stars located within or near to the interstellar cloud to light it up. The Coalsack is also quite large, with parts of it crossing the constellation's borders into neighboring Centaurus and Musca.

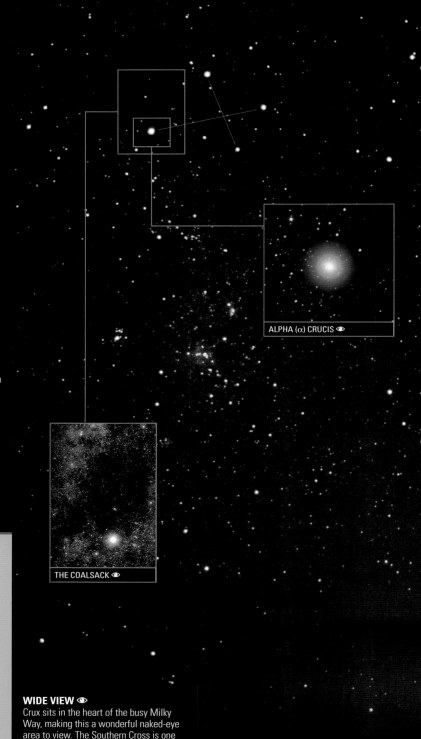

ALPHA (α) CRUCIS 👁

THE COALSACK 👁

WIDE VIEW 👁
Crux sits in the heart of the busy Milky Way, making this a wonderful naked-eye area to view. The Southern Cross is one of the most famous celestial patterns.

CRUX (Cru)	
WIDTH / DEPTH	✋ / ✋
SIZE RANKING	88th
FULLY VISIBLE	25°N–90°S

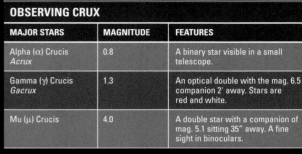

OBSERVING CRUX

MAJOR STARS	MAGNITUDE	FEATURES
Alpha (α) Crucis *Acrux*	0.8	A binary star visible in a small telescope.
Gamma (γ) Crucis *Gacrux*	1.3	An optical double with the mag. 6.5 companion 2' away. Stars are red and white.
Mu (μ) Crucis	4.0	A double star with a companion of mag. 5.1 sitting 35" away. A fine sight in binoculars.

NOTABLE OBJECTS	MAGNITUDE	FEATURES
NGC 4755, the Jewel Box	4.2	An open star cluster that sits around the star Kappa (κ) Crucis and is around 6,500 light-years distant.
The Coalsack	—	A dark nebula; to the eye, this appears as an unmistakable "hole" in the Milky Way.

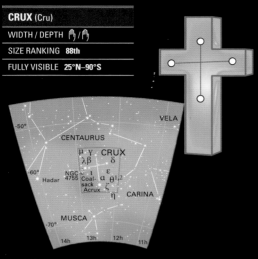

SKY MAP
The constellation of Crux lies between the "legs" of neighboring Centaurus. Known to the ancient Greeks, it is also important in Aboriginal and Maori mythology.

THE FALSE CROSS 👁
The nearby False Cross asterism can be mistaken for the Southern Cross constellation, which in turn can lead to errors in navigation. The False Cross (to the right in the image) is formed from two stars in Vela and two in Carina. Crux (on the left of the image) is distinguishable by the presence of a fifth star—Epsilon (ε) Crucis—almost in line between Alpha (α) and Delta (δ) Crucis.

GAMMA (γ) CRUCIS
At the top of Crux lies the 2nd-magnitude red giant Gamma (γ) Crucis, or Gacrux, around 88 light-years from Earth. Binoculars show an unrelated companion, six times more distant.

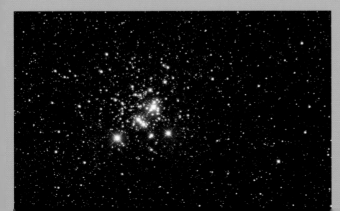

NGC 4755, the Jewel Box 👁
The Jewel Box cluster is a small but wonderful open star cluster that to the naked eye resembles a single fuzzy star. In binoculars, its many bright, multicolored stars are a fantastic sight, and the Jewel Box is widely regarded as one of the finest open star clusters in the night sky. Through its center hangs a chain of three close stars, the northernmost of which

TREASURE CHEST
Sitting just to the north of the Coalsack is NGC 4755, the wonderful Jewel Box cluster of around 50 stars, seen here with the aid of a small telescope.

is a red-tinged complement to the other two blue stars. Despite its apparent proximity, the Jewel Box is 10 times further from Earth than the Coalsack, which is only 2,000 light-years away

STARHOPPING FROM CRUX

Crux is the southern hemisphere's equivalent of the Big Dipper: a prominent pattern that can be used to find the sky's south pole and many other constellations.

The Southern Cross, or Crux, is the smallest but one of the best-known of all the constellations. Lying in a rich area of the Milky Way, it is defined by just four stars in a cross formation. Its long axis, which points the way to the south celestial pole, is crossed by a shorter horizontal "beam."

1 TO CENTAURUS

Start at Delta (δ) Crucis, the fainter of the two stars on the beam of Crux. Extend a line through Beta (β) Crucis, or Mimosa, the other star on the beam. You'll soon reach the bright (magnitude 0.6) star Hadar, Beta (β) Centauri, and then the equally bright Rigil Kentaurus, Alpha (α) Centauri—the two leading stars of Centaurus.

CENTAURUS

2 TO OCTANS AND THE POLE

There is no bright star at the position of the south celestial pole, though the faint constellation Octans lies nearby. To find the pole, extend a line through the two leading stars in Centaurus (see above). At the mid-point of this line, draw another line at right-angles. Then, extend another line from the long axis of Crux (from Gacrux, Gamma (γ) Crucis to Acrux, Alpha (α) Crucis) . Where these two lines meet marks the southern pole.

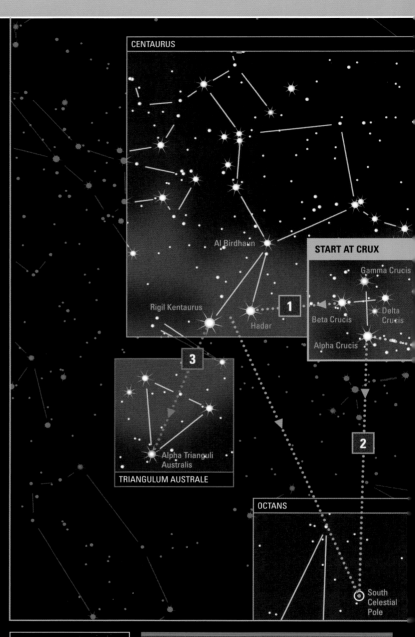

CENTAURUS

Al Birdhaun

Rigil Kentaurus

Hadar

START AT CRUX

Gamma Crucis

1

Beta Crucis

Delta Crucis

Alpha Crucis

3

Alpha Trianguli Australis

TRIANGULUM AUSTRALE

2

OCTANS

South Celestial Pole

TRIANGULUM AUSTRALE

3 TO TRIANGULUM AUSTRALE

Extend a line from Al Birdhaun, or Epsilon (ε) Centauri, through Rigil Kentaurus, its neighboring star in the constellation of Centaurus. The next fairly bright star (magnitude 1.9) that you encounter will be Atria, Alpha (α) Trianguli Australis, which is the apex of this triangular constellation. The other two vertices of this almost-equilateral triangle sit equally spaced in the direction back toward Centaurus.

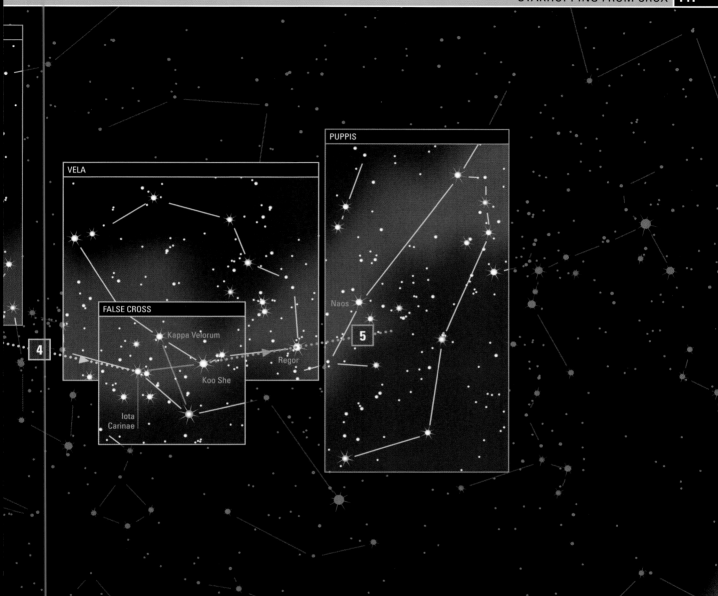

VELA

PUPPIS

FALSE CROSS

Kappa Velorum

Koo She

Iota Carinae

Regor

Naos

4

5

4 TO THE FALSE CROSS

There is another cross shape in the southern sky which inexperienced stargazers may confuse with Crux. This asterism, known as the False Cross, can be found by starhopping from Hadar in Centaurus. Draw an imaginary line from here through Alpha Crucis, and extend it over the Milky Way. At about twice the distance between these first two stars, one of the stars on the False Cross's "beam," Iota (ι) Carinae, will be reached.

PUPPIS

5 TO VELA THEN PUPPIS

The two stars that make up the "beam" of the False Cross are Iota (ι) Carinae (from Carina) and Koo She, Delta (δ) Velorum (from Vela). By extending a line between these two stars in the opposite direction to Crux, you will reach the leading star of Vela, the Sail, known as Regor. Continuing this line takes you out of Vela and into the constellation of Puppis, the Stern, with its brightest star, Naos, or Zeta (ζ) Puppis.

CENTAURUS

Centaurus, the Centaur, is home to our nearest star neighbor and the superb globular cluster Omega Centauri.

Centaurus is a large constellation with a lot to offer the amateur astronomer. The four "feet" of the mythological Greek creature are firmly embedded in a splendidly bright and complex part of the Milky Way. It boasts two magnificent sights: the open star cluster sitting across the nebula IC 2944 and the famous globular cluster Omega Centauri. In addition, leading the way are the two brilliant stars Alpha (α) Centauri, also known as Rigil Kentaurus, and Beta (β) Centauri, or Hadar. Although appearing close in the sky, in reality the vast distances of space come into play, for while Hadar is around 320 light-years from Earth, Rigil Kentaurus is only just over 4 light-years away—it is, in fact, the closest star system to us after the Sun. The closest individual star in this system is Proxima Centauri, an 11th-magnitude red dwarf that needs a telescope to be seen.

Centaurus is home to Proxima Centauri, the nearest star to our solar system.

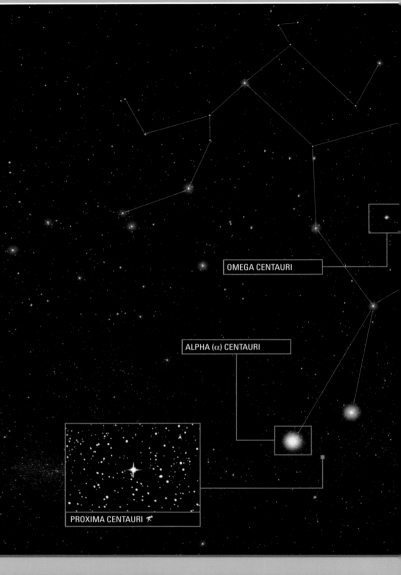

OMEGA CENTAURI

ALPHA (α) CENTAURI

PROXIMA CENTAURI ✈

NGC 5139, Omega Centauri 👁

To the eye, this starlike object looks as though it has been slightly smudged in the sky. However, with ever-increasing sizes of telescope, the view of Omega Centauri becomes more and more impressive, as larger numbers of its estimated one million stars are revealed. It is the largest globular cluster of the Milky Way, up to 10 times as massive as other globular clusters. It is also one of the closest to us at only around 17,000 light-years from the Solar System—so Omega Centauri is also the brightest globular cluster in the night sky. Studies of the cluster have shown that Omega Centauri is one of the oldest objects in the Milky Way—at 12 billion years old it is almost as ancient as the Universe itself.

A MILLION-STAR CLUSTER
To the eye, Omega Centauri appears as a fuzzy star, but a small telescope will start to clearly distinguish its elliptical shape (left), and larger ones will resolve its individual stars (above).

IC 2944 👁

CENTAURUS (Cen)

WIDTH / DEPTH	🖐🖐🖐 / 🖐🖐
SIZE RANKING	9th
FULLY VISIBLE	25°N–90°S

SKY MAP
Using the chart, it is easier to see how the stars can be joined to resemble the outline of the classical Centaur.

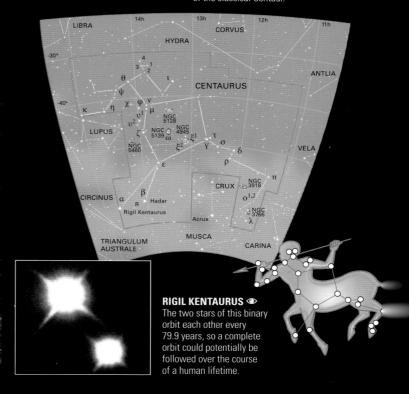

RIGIL KENTAURUS 👁
The two stars of this binary orbit each other every 79.9 years, so a complete orbit could potentially be followed over the course of a human lifetime.

WIDE VIEW 👁
The main stars of the constellation stand out well: the bright, yellowish Alpha (Rigil) on the left and the almost equally bright Beta (Hadar) to its right.

Proxima Centauri 🏃

This 11th-magnitude red dwarf was only discovered in 1915. It is not a stunning sight when viewed through a telescope, but the satisfaction comes from knowing you are viewing the closest star to the Sun, only 4.2 light-years distant. It is thought to be part of the Alpha (α) Centauri system, although its orbital period is around one million years, leading some astronomers to question whether Proxima is gravitationally bound to Alpha (α) Centauri at all. Due to its closeness to Earth, some scientists have investigated whether there could be habitable planets around Proxima. However, as it is a cool red dwarf—and a flare star, prone to radiation outbursts—the chances of finding such a planet in the system must be very slim.

OBSERVING CENTAURUS

MAJOR STARS	MAGNITUDE	FEATURES
Alpha (α) Centauri *Rigil Kentaurus*	-0.3	A bright binary star; the main mag. -0.01 star is accompanied by a smaller mag. 1.4 companion.
R Centauri	5.3–11.8	A Mira-type long-period variable star whose brightness changes over 546 days.
Proxima Centauri	11.1	The closest star to the Sun, just 4.2 light-years away.

NOTABLE OBJECTS	MAGNITUDE	FEATURES
NGC 3766	5.3	An open star cluster, visible to the eye; a telescope is needed to start resolving the 100 or so stars.
NGC 3918, the Blue Planetary	8.5	A planetary nebula visible with small telescopes. It appears like a rounded blue disc, similar to Uranus.
NGC 5139, Omega (ω) Centauri	3.7	A fantastic 12 billion-year-old globular cluster that sits around 17,000 light-years away from us.
IC 2944	5.5	A nebula near Lamda (λ) Centauri, lying behind an alignment of four stars that stretch across its center.

CARINA

Part of Carina, the Keel, is dominated by the Milky Way band. Its other half may be less busy, but it contains the intensely bright leading star, Canopus (Alpha Carinae).

What Carina lacks in a recognizable pattern, it certainly makes up for with a wonderful variety of deep-sky objects—including Canopus, the second-brightest star in the sky. The deep-sky objects cannot be seen with the naked eye, but are easily revealed with binoculars. Once the French astronomer Nicolas de Lacaille broke up the ancient Greek constellation Argo Navis (the Ship of the Argonauts), in the 18th century, its component parts—Carina, Vela (the Sails), and Puppis (the Stern)—were left rather shapeless. The one exception to this is an asterism of four stars called the False Cross, which is often the source of navigational confusion due to its proximity and similarity to Crux, the Southern Cross (see pp.138–139). This asterism is created by the stars Epsilon (ε) and Iota (ι) Carinae, and, from the neighboring constellation of Vela, Delta (δ) and Kappa (κ) Velorum. Although this group looks uncannily similar to Crux, it does not all lie within the Milky Way but on the band's edge. The False Cross is larger than Crux, but its stars are not so colorful nor as bright.

NGC 2516 👁

Discovered in 1751–1752 by the astronomer Nicolas de Lacaille, NGC 2516 is a bright open star cluster that contains around 100 stars. It lies around 1,300 light-years away from Earth, and is visible with the naked eye, with an apparent size equal to the full Moon in the night sky. Binoculars reveal a fine cross pattern to its stars, as well as several colorful stars—always a bonus for deep-sky objects—the brightest being a 5th-magnitude red giant. It appears as though there are at least three spiralling chains of stars that emanate from NGC 2516's center, leading to the popular name for this cluster: the Garden Sprinkler. It is also known as the Diamond Cluster due to the notable clarity of its stars.

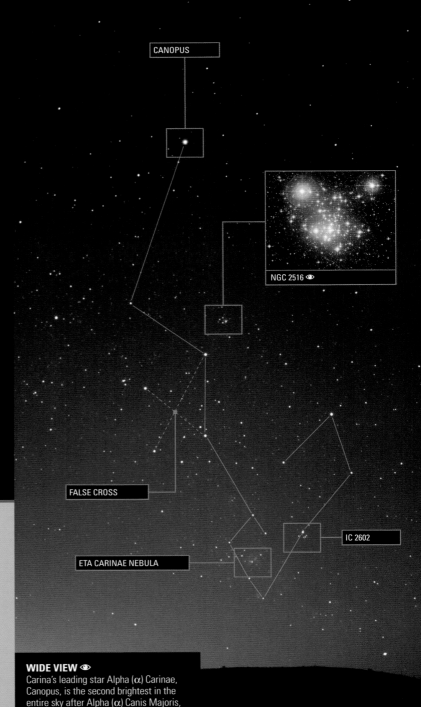

CANOPUS

NGC 2516 👁

FALSE CROSS

IC 2602

ETA CARINAE NEBULA

WIDE VIEW 👁
Carina's leading star Alpha (α) Carinae, Canopus, is the second brightest in the entire sky after Alpha (α) Canis Majoris, Sirius (the Dog Star), in Canis Major.

CARINA (Car)

WIDTH / DEPTH	✋✋ / ✋
SIZE RANKING	**34th**
FULLY VISIBLE	**14°N–90°S**

IC 2602 👁

This cluster, based around the star Theta (θ) Carinae, is visible to the naked eye, but it makes for a finer sight in binoculars.

OBSERVING CARINA

MAJOR STARS	MAGNITUDE	FEATURES
Alpha (α) Carinae *Canopus*	-0.7	A white star that sits 310 light-years away from the Solar System.
Eta (η) Carinae *Foramen*	6.2	An irregular variable star that reached mag. -0.8 in April 1843. It is around 7,000 light-years from Earth.
R Carinae	3.9–10.5	A Mira-type variable star with a period of about 308 days.

NOTABLE OBJECTS	MAGNITUDE	FEATURES
NGC 2516, the Garden Sprinkler Cluster	3.8	An open star cluster, so-named because of its spiral "sprinkler" appearance.
NGC 2808	6.3	One of our galaxy's massive globular clusters, home to over a million stars.
Eta Carinae Nebula, NGC 3372	1.0	An emission nebula that appears as a hazy orange ellipse through a small telescope.
IC 2602, the Southern Pleiades	1.9	Binoculars will provide a great view of this open star cluster, 479 light-years away.

SKY MAP

The northern part of Carina, near the border with Crux, the Southern Cross, is filled with star clusters and nebulae that can easily be seen in binoculars.

Carina's **leading** star, **Canopus**, is the **second-brightest** star in the night sky.

The Eta Carinae Nebula 👁

The constellation's undoubted centerpiece is the Eta Carinae Nebula. Embedded and formed within this gaseous realm are many young stars, one of which is the massive Eta (η) Carinae. Recent studies show that this is a multiple system, but the largest star is the most noticeable, believed to be over 100 times more massive than the Sun and 4 million times brighter. In fact, Eta (η) Carinae is only just able to keep itself together by gravity, as its radiation force is almost ripping the star apart. It has produced brightening explosions over the years. One, in 1843, saw the star briefly become the second brightest in the night sky.

STELLAR NURSERY

An emission nebula 7,500 light-years from the Solar System, the Eta Carinae Nebula is one of the great star-forming regions in the Milky Way.

GLOWING GAS CLOUDS
The Eta Carinae Nebula is a vast patch of glowing gas that covers an area of night sky four times as wide as the full Moon. It is visible to the naked eye against the backdrop of the Milky Way.

VELA

The two southernmost stars of Vela, the Sails, combined with two others from neighboring Carina, create the False Cross asterism (see p.144). Finding this is the best way to locate Vela.

Vela, the Sails, was originally part of the larger constellation of Argo Navis, the Ship, whose division left its three component constellations, Vela, Puppis (the Stern), and Carina (the Keel), with patternless appearances. It also meant that some of their star designations suffered. For example, Vela has no alpha (α) or beta (β) stars, so its leading star is Gamma (γ) Velorum (Regor), which sits on the edge of the constellation. Vela is the home of NGC 2547, a family of about 50 stars visible to the eye, but with binoculars a north–south meander of stars can be seen through its center. Vela also contains one of the largest residues of an exploded star, the Vela Supernova Remnant, wispy traces of which may be caught by larger amateur telescopes.

The Milky Way's path through Vela offers a fine array of open star clusters.

NGC 3132, the Eight-Burst Nebula ✈
The Eight-Burst Nebula is so-named because of its loops of gas that interlock like figures-of-eight. These are only revealed through a large telescope or on long-exposure photographs. NGC 3132 is also referred to as the Southern Ring Nebula, due to its similarity in appearance to the northern hemisphere's Ring Nebula (M57), in Lyra. Its oval form is easily viewed in a small telescope, which may also reveal the 10th-magnitude star at its center. There are many things about this planetary nebula that are not known, which has made NGC 3132 an object of academic study. Unanswered questions include why its overall shape is so unsymmetrical, and how the dust lanes across its center are formed.

REGOR

NGC 2547 ♔

NGC 3132, THE EIGHT-BURST NEBULA ✈

WIDE VIEW 👁
Vela contains a few bright stars, but they are not linked in any recognizable way, having been separated from the original, much larger constellation of Argo Navis.

VELA (Vel)

WIDTH / DEPTH	
SIZE RANKING	**32nd**
FULLY VISIBLE	**32°N–90°S**

SKY MAP

The Milky Way makes Vela a worthwhile constellation to scan with binoculars, as many stars and clusters will become apparent.

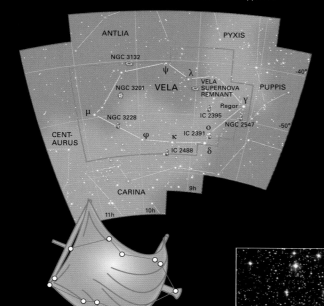

OBSERVING VELA

MAJOR STARS	MAGNITUDE	FEATURES
Gamma (γ) Velorum *Regor*	1.8	An optical double star with a mag. 4.3 companion that orbits it every 78.5 days.

NOTABLE OBJECTS	MAGNITUDE	FEATURES
NGC 2547	4.7	An open star cluster: a nice arrangement of stars can be seen through binoculars.
NGC 3132, the Eight-Burst Nebula	8.1	A planetary nebula whose oval shape and central star are quite easy to find with a small telescope.
NGC 3201	8.2	A globular cluster that can be seen through binoculars, but a telescope is needed to resolve any of its stars.
IC 2391	2.5	An open star cluster of 30 stars sitting around 580 light-years away.

RED GIANT CLUSTER

The globular cluster NGC 3201 contains many bright red giant stars (right), giving it an overall reddish appearance. It is too faint to be seen with the naked eye, however, with a visual magnitude of only 8.2, and appears to be less condensed than most globular clusters.

The Vela Supernova Remnant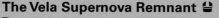

There are various supernova remnants dotted around the sky that help us journey back a long way in time. From their size and distance, it is often possible to tell how long ago the star exploded. The Vela remnant, between Gamma (γ) and Lambda (λ) Velorum, was created from an exploding star around 12,000 years ago. Matter from the star has now been spread over 55 light-years through space, and as the remnant is only about 800 light-years away, this means it covers an area of sky ten times the apparent size of the full Moon. The filaments of gas and dust scattered over this area can be seen by powerful telescopes from dark sites.

A BIG BANG
The explosion of the star that produced the Vela Supernova Remnant (above) may have rivaled the brightness of the Moon. NGC 2736, the Pencil Nebula (right), is a relic of this explosion.

DORADO

Dorado, the Goldfish, would be an unimportant constellation were it not for the presence of the naked-eye marvel that is the Large Magellanic Cloud.

Lying mostly within Dorado is a dwarf galaxy close to our own. This galaxy, the Large Magellanic Cloud (LMC), is named after the 16th-century explorer Ferdinand Magellan, who brought it to the attention of the northern hemisphere—where, for many, Dorado is not visible. The LMC is the biggest of two such starry islands that are probably satellites of the Milky Way and look almost like detached parts of our galaxy. Binoculars or a small telescope will reveal many open star clusters, globular star clusters, and nebulae. One bright object observed within the LMC was originally designated as a star, "30 Doradus." However, this was revealed to be something entirely different when an optical aid was trained upon it. 30 Doradus is now known to be the Tarantula Nebula, a vast and stunning bright cloud that sits on the edge of the LMC.

The **LMC** is a **dwarf** galaxy lying around **179,000** light-years from the **Milky Way**.

The Large Magellanic Cloud (LMC) 👁

Our knowledge of the Large Magellanic Cloud has evolved as astronomers' techniques and instruments have improved. Once believed to be the closest external galaxy to our own, at 179,000 light-years from the Solar System, it is now at third place after the discovery of the Canis Major Dwarf and the Sagittarius Dwarf elliptical galaxies. Also, although it was originally classified as an irregular galaxy, surveys now indicate that it has a starry barlike feature across its center, from which a weak spiral structure appears. Overall, the LMC is about one-tenth the size of our galaxy, making it the fourth-largest galaxy in the Local Group. It appears to us around 20 times the size of the full Moon.

BOLE

WIDE VIEW 👁
The pattern of stars forming Dorado starts north of the LMC. The "Goldfish" is imagined to be swimming toward the south celestial pole.

LARGE MAGELLANIC CLOUD 👁

DORADO (Dor)

WIDTH / DEPTH	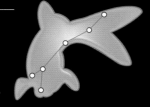
SIZE RANKING	72nd
FULLY VISIBLE	20°N–90°S

SKY MAP
Dorado is mostly surrounded by equally faint constellations, such as Horologium, the Clock, and Caelum, the Chisel.

OBSERVING DORADO

MAJOR STARS	MAGNITUDE	FEATURES
Alpha (α) Doradus *Bole*	3.3	A relatively faint star that is still the brightest in Dorado; it is a binary star, composed of a smaller B star revolving around a giant A star.
Beta (β) Doradus	3.5–4.1	One of the brightest Cepheid variable stars with a period of 9.8 days.
R Doradus	4.8–6.6	A red giant Mira-type long-period variable with a semi-regular period of about 11 months.

NOTABLE OBJECTS	MAGNITUDE	FEATURES
Large Magellanic Cloud (LMC)	0.4	An irregular dwarf galaxy lying 179,000 light-years away; a fine sight in binoculars or a small telescope.
NGC 2070, the Tarantula Nebula	5.0	An emission nebula so large that if it was as near to us as the Orion Nebula, it would cover half the sky.

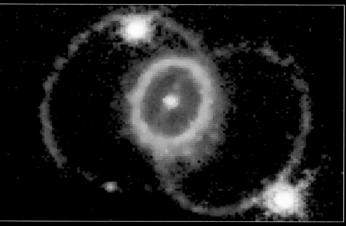

SUPERNOVA 1987A
The first bright supernova since the invention of the telescope occurred in the LMC in February 1987. Shown here in a Hubble Space Telescope image (right), it was visible to the naked eye for two months.

NGC 2070, the Tarantula Nebula 👁

The magnificent Tarantula Nebula lies off to one side of the main stellar bar of the LMC. Its brightness and size can be appreciated still further by comparing it to the famed Orion Nebula (M42), which is in our galaxy and sits only 1,350 light-years away. The Tarantula is located in another galaxy altogether, about 160,000 light-years away. The fact it is visible to the eye at such an incredible distance means it is a truly bright and enormous object—more than 20 times as big, in fact, as M42. Its long, tarantulalike arms of hydrogen may stretch up to 1,800 light-years from the nebula's bright center, where its main star-forming region is located.

THE TARANTULA NEBULA
The nebula's gas arms give it a spidery shape. It appears to the naked eye as a large milky patch, around the same size as the full Moon in the sky.

TUCANA

One of the Milky Way's lesser galactic companions, the Small Magellanic Cloud, sits within the boundaries of Tucana, the Toucan.

The southern hemisphere has many examples of celestial objects that, in pre-telescope days, were designated as single stars—only later to be revealed as vast spherical stellar islands known as globular clusters. This happened in the case of 47 Tucanae, or NGC104. Using a small telescope, or even binoculars, it is hard to choose whether 47 Tucanae or Omega (ω) Centauri in Centaurus is the sky's most impressive globular, as both offer spectacular views. 47 Tucanae sits in our galaxy, appearing to lie near the constellation's main naked-eye feature, the Small Magellanic Cloud (SMC), a satellite galaxy of the Milky Way. Like the Large Magellanic Cloud in Dorado, the SMC looks as though a piece of our galaxy has somehow drifted off into a darker region of the sky, and binoculars will start to reveal its star clusters and nebulae. Tucana also boasts a smaller globular cluster, NGC 362.

To the **naked eye**, the globular cluster **47 Tucanae** appears to be a **hazy** 4th-magnitude **star**.

The Small Magellanic Cloud (SMC) 👁

Lying around 200,000 light-years away, the SMC is an irregular dwarf galaxy orbiting the Milky Way, but it is still the fifth-largest of all the galaxies in the Local Group. This is a family of about 30 galaxies, the largest of which are our own galaxy and the Andromeda Galaxy (M31). Between these two is the Group's center of gravity. The only other large member of the Local Group is M33, in Triangulum, while the rest, like the SMC, are dwarf galaxies. The SMC forms a wedge-shaped cloud that looks like a detached part of the Milky Way, and is around seven times wider than the apparent size of the full Moon. Binoculars or a small telescope can distinguish individual clusters and nebulae within it.

BETA (β) TUCANAE

SMALL MAGELLANIC CLOUD 👁

WIDE VIEW 👁
Tucana is not a bright constellation, but it can be identified by finding a crooked line of three stars—seen here to the right of the SMC—starting with Beta (β) Tucanae.

TUCANA (Tuc)

WIDTH / DEPTH	🖐️🖐️ / 🖐️
SIZE RANKING	**48th**
FULLY VISIBLE	**14°N–90°S**

SKY MAP
Tucana is found in a quiet area of the sky, with the exception of the SMC and the bright star Achernar, in Eridanus, which sits just outside Tucana.

NGC 362 🔭
The second globular cluster in Tucana, NGC 362 (right) is smaller and fainter than 47 Tucanae, and requires binoculars or a small telescope to be seen. It is a foreground object in our own galaxy, but in the night sky it appears to be lying by the northern tip of the SMC.

OBSERVING TUCANA

MAJOR STARS	MAGNITUDE	FEATURES
Beta (β) Tucanae	4.4	A system of six stars; the brightest forms a double star with a mag. 4.5 companion.
Kappa (κ) Tucanae	5.1	A double star with a mag. 7.3 companion 5' 30" away.

NOTABLE OBJECTS	MAGNITUDE	FEATURES
Small Magellanic Cloud (SMC)	2.3	An irregular dwarf galaxy 200,000 light-years from the Milky Way; its central region is cataloged as NGC 292.
NGC 104 47 Tucanae	4.0	A compact globular cluster, considered to be one of the most spectacular of its kind in the night sky.

47 Tucanae 👁️

47 Tucanae, formally listed as deep-sky object NGC 104, was originally cataloged as a star. It was the 47th in order of right ascension in Tucanae, but was later discovered to be the second-largest and second-brightest globular cluster in the sky. A telescope will show it to be an example of the more compact type of globular cluster, with a brighter center. It has quite a large visual size in the sky too, with an apparent diameter equal to that of the full Moon. In real-space terms, sitting as it does 13,400 light-years from the Solar System, 47 Tucanae equates to a cluster that is 120 light-years in diameter.

CROWDED CLUSTER
47 Tucanae (above) contains several million stars—enough to fill a small galaxy. The cluster's central region is so crowded (right) that there is a high rate of stellar collision.

PAVO

The main feature of Pavo, the Peacock, is its leading star, Alpha (α) Pavonis. This has also been given the name "Peacock," a star name first used in an English almanac in the 1930s.

Pavo sits on the fringes of the Milky Way band and has a high declination, meaning that it appears close to the south celestial pole. Without the Milky Way's starry band to obstruct the view, stargazers are able to see further into deep space. At Pavo, this allows a telescopic view of the great galaxy, NGC 6744, as well as the third-largest globular cluster of the skies, NGC 6752. To the naked eye, there is little pattern to Pavo's main stars, but a close look in dark skies will show some faint star groups. Some are doubles, like the great Mu (μ) Pavonis, or Phi (φ) Pavonis, which makes a trio with nearby Rho (ρ). At the top of the Peacock's "tail" is Theta (θ) Pavonis, one of a chain of three that resemble a smaller version of Orion's Belt.

At **number 44,** Pavo is exactly **midway** in the **size rankings** of the 88 constellations.

NGC 6744 ✸

PEACOCK

NGC 6744 ✸
An impressive barred-spiral galaxy, presented virtually face-on to Earth, NGC 6744 can be seen through a small or medium telescope as an elliptical haze sitting in a rich starfield. However, larger instruments will be needed to clearly identify its spiral arms, and only photographs can reveal what a stunning object this galaxy, 25–30 million light-years distant, truly is. Studies suggest that if seen from outside, the Milky Way would resemble NGC 6744, with its loose arrangement of arms and a similar bar running through its large, elongated core. NGC 6744 also has at least one smaller companion galaxy, which appears superficially similar to the Large and Small Magellanic Clouds.

WIDE VIEW 👁
Pavo is part of a large, indistinct area of the far-southern sky, with just a few odd bright stars to break up the emptiness. It is located on the edge of the Milky Way.

PAVO (Pav)

WIDTH / DEPTH	🖐🖐 / 🖐
SIZE RANKING	**44th**
FULLY VISIBLE	**15°N–90°S**

SKY MAP
Pavo lies just north of Octans, the even fainter constellation that is the home of the south celestial pole.

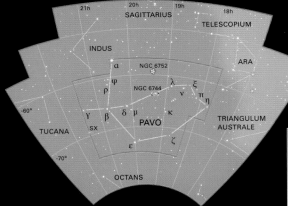

OBSERVING PAVO

MAJOR STARS	MAGNITUDE	FEATURES
Alpha (α) Pavonis *Peacock*	1.9	A bright double star in the Peacock's "throat."
Kappa (κ) Pavonis	3.9–4.8	A bright Cepheid variable star with a period of 9 days.
Mu (μ) Pavonis	5.3	A double star with a mag. 5.7 partner. Both stars are orange.

NOTABLE OBJECTS	MAGNITUDE	FEATURES
NGC 6744	9.2	A barred-spiral galaxy with a large apparent size in the sky.
NGC 6752	5.4	A globular cluster that is great to view with a small telescope, which will help distinguish its stars.

THE PEACOCK IN MYTHOLOGY
In Greek mythology, the peacock was the sacred bird of Hera, wife of Zeus. She traveled on a chariot pulled by peacocks, as seen in this painting by Rubens, *The Birth of the Milky Way* (1636–37). It depicts the creation of the galaxy from milk spilled by Hera as she breast-fed the hero Hercules. The word "galaxy" comes from the ancient Greek *galaxias*, or "milky."

NGC 6752 📷
This globular cluster is one of the best to observe. It is very easy to find with binoculars due to its brightness and its apparent size, which is half that of the full Moon. In visual size, NGC 6752 is only beaten by Omega (ω) Centauri in Centaurus and 47 Tucanae in Tucana. The view is enhanced by an unrelated foreground star of magnitude 7.6 that sits to the south of the central cluster. Observation with the eye, of course, gives us no information about objects' relative distances from us. Here, the foreground star, actually a close binary, is 788 light-years from the Solar System, while the globular cluster itself is around 15,000 light-years distant.

SOUTHERN CLUSTER
The fine globular cluster NGC 6752 remains relatively unstudied because of its far-south declination. It lies at the limit of naked-eye visibility.

LIBRA AND HYDRA

Libra, the Scales, is most famous for being one of the twelve constellations of the zodiac, and Hydra, the Water Snake, is the largest constellation of them all.

The main features of Libra—the scales of justice—are its stars. Here there are some fine doubles and some even finer names: Alpha (α) Librae and Beta (β) Librae are called respectively Zubenelgenubi and Zubeneschamali, translated as the Southern and Northern Claws of the Scorpion. This refers to a time when the ancient Greeks considered Libra to be a part of Scorpius. Libra can be used to locate Hydra, which it borders. Hydra represents the multi-headed serpent slain by the Greek hero Hercules, and is truly an enormous constellation, spanning almost one-third of the entire sky. However, despite this huge range it does not meet the Milky Way once, and with only one brightish star, Alpha (α) Hydrae—at magnitude 2.0, the constellation is not as spectacular to the naked eye as you might expect. The tail of the Water Snake begins immediately to the right of the star Sigma (σ) Librae. As the constellation weaves toward Monoceros, the dedicated stargazer is rewarded with a fine variable star (R Hydrae), the M83 galaxy, a bright planetary nebula—the Ghost of Jupiter—and the M48 open cluster.

M48 👁

This open star cluster, 1,500 light-years distant from Earth, lies very near Hydra's border with Monoceros, and is estimated to be around 300 million years old. M48 is a loose collection of at least 80 stars that can be seen with the eye as a large, faint fuzzy patch, with an apparent size larger than that of the full Moon. It is best observed through binoculars or a small telescope, because then you can see that the cluster contains a few orange and yellow stars mixed in with the group. M48 was discovered by the French astronomer Charles Messier in 1771, but due to an error in calculation its position was incorrectly entered into his catalog, and the object was not found again until 1783.

LIBRA

ANTARES

ZUBENELGENUBI

WIDE VIEW 👁
Libra can be found easily by using Antares and other nearby stars in neighboring Scorpius, which point the way to where this constellation is located.

HYDRA

M48

M83 ✶

WIDE VIEW 👁
To the naked-eye observer, the most identifiable parts of Hydra are its head and eye, formed from six stars located just south of Cancer (see Sky Map).

LIBRA (Lib)

WIDTH / DEPTH	🖐️ / 🖐️
SIZE RANKING	29th
FULLY VISIBLE	60°N–90°S

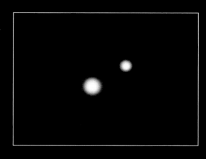

SKY MAP: LIBRA

The ecliptic runs right through the heart of Libra, so planets can occasionally brighten this recognizable but not outstanding constellation.

OBSERVING LIBRA

MAJOR STARS	MAGNITUDE	FEATURES OF INTEREST
Alpha (α) Librae *Zubenelgenubi*	2.8	A naked-eye double star—known as the Southern Claw—with the smaller companion at mag. 5.2. The stars are bluish and yellow in color.
Delta (δ) Librae *Zubenelakribi*	4.9–5.9	A fairly bright Algol-type eclipsing binary star which dips in magnitude every 2 days 8 hours.

ALPHA (α) LIBRAE 👁️

Zubenelgenubi is a naked-eye, wide, double star, lying around 77 light-years from the Solar System. It is a good object to observe in darker skies, with the two stars—appearing at magnitude 2.8 and 5.2—separated with binoculars. It lies close to the ecliptic and can be occulted by the Moon and (rarely) the planets.

OBSERVING HYDRA

MAJOR STARS	MAGNITUDE	FEATURES OF INTEREST
54 Hydrae	5.3	A double star with a mag. 7.4 companion, near the border with Libra. Its stars are yellow and purple in a telescope.
R Hydrae	4.5–9.5	A Mira-type variable with a period of around 389 days.

NOTABLE OBJECTS	MAGNITUDE	FEATURES OF INTEREST
M48	5.8	A visually large open star cluster sitting about 1,500 light-years away from Earth.
M83	7.5	The "Southern Pinwheel" spiral galaxy, head-on to Earth.
NGC 3242	7.7	A planetary nebula, also known as the Eye Nebula or the Ghost of Jupiter.

HYDRA (Hya)

WIDTH / DEPTH	🖐️🖐️🖐️🖐️🖐️ / 🖐️🖐️
SIZE RANKING	1st
FULLY VISIBLE	54°N–83°S

SKY MAP: HYDRA

Hydra covers 1,303 square degrees of the sky. This makes it the largest constellation, beating Virgo into second place by 9 square degrees.

PUPPIS

Puppis, the Stern, has some fine open star clusters, such as M46 and M47, which are close enough in the sky to be seen together in binoculars.

Puppis is the largest part of the old constellation Argo Navis, which was broken up by the astronomer Nicolas de Lacaille in 1763. This created three new constellations, all rather bereft of shape, and so it is hard to recognize the pattern of stars. In addition, because the stars were not redesignated, there is no "Alpha (α) Puppis" here; the leading star is Zeta (ζ) Puppis, or Naos (Greek for "ship"). It is a blue supergiant with a surface temperature of around 75,600°F (42,000°C), making it the hottest star visible to the naked eye. Naos sits near the border with Vela, lying just within the Milky Way band. When viewing here you are beginning to move out of the galactic plane, so the Milky Way is not as bright as it appears in Sagittarius or Scorpius, where you are looking into the center of our galaxy.

Puppis is close to a line that joins the sky's two brightest stars, Sirius and Canopus.

M46 and M47 👁

To the naked eye, the prominent open star cluster M46 appears to be merely a brighter patch of the Milky Way band. It is a close group of stars, over 5,000 light-years from Earth, that are all rather similar in brightness. However, any observation will be greatly enhanced by the use of a telescope, which will distinguish individual stars. Lurking in the cluster's foreground is NGC 2438. This is a planetary nebula of magnitude 10.8 that is almost twice as distant from us as M46, and its visual proximity is just one of those lucky positional coincidences. M46 forms a pair of star clusters with the nearby M47, which is closer to Earth (at 1,500 light-years' distance) and is the brighter of the two.

SIRIUS

M47 👁

M46 👁

WIDE VIEW 👁
Puppis has a fine selection of little starry groups that are visible to the naked eye, and are a wonderful sight to observe through binoculars.

PUPPIS (Pup)

WIDTH / DEPTH	
SIZE RANKING	20th
FULLY VISIBLE	39°N–90°S

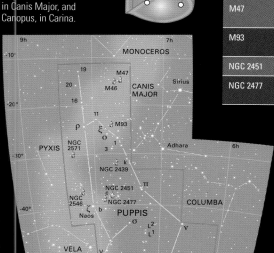

SKY MAP
The constellation of Puppis can be located close to a line that joins the sky's brightest two stars: Sirius, in Canis Major, and Canopus, in Carina.

OBSERVING PUPPIS

MAJOR STARS	MAGNITUDE	FEATURES
Kappa (κ) Puppis	4.5	A fine double star with a mag. 4.6 companion 10" away.
Zeta (ζ) Puppis *Naos*	2.2	The leading star of Puppis, and a blue supergiant whose mass is almost 60 times greater than that of the Sun.
L1 Puppis	4.9	An optical double star with L2 Puppis.
V Puppis	4.4–4.9	An eclipsing binary star that varies in magnitude over 1.5 days.

NOTABLE OBJECTS	MAGNITUDE	FEATURES
M46	6.1	A 5,400 light-year-distant open star cluster with an unrelated planetary nebula appearing within it.
M47	4.4	An open star cluster: binoculars will reveal the loose association of this family of 40 stars.
M93	6.2	A small but bright open star cluster that superficially resembles a squashed Z-shape.
NGC 2451	2.8	An open star cluster, with a close family of stars, 850 light-years away.
NGC 2477	5.7	An open star cluster with the appearance of a weak globular cluster, that is wonderful to view through a small telescope.

SMALL STAR GROUP
A bright open star cluster, M93 is relatively small in celestial terms, at around 25 light-years across. It is 100 million years old, which is young by astronomical standards. M93 lies near the celestial equator, and consists of around 80 stars.

NGC 2477
With more than 300 stars in a roundish group, NGC 2477 is one of the finest open star clusters in Puppis. Its apparent size in the sky is just smaller than that of the full Moon, and overall it is so rich in stars that it seems more like a weak globular cluster, with some of the stars appearing in curving arcs. NGC 2477 lies around 4,200 light-years from the Solar System, and is located on a line between Zeta (ζ) Puppis (Naos) and Pi (π) Puppis (Ahadi). It was discovered in 1751 by the French astronomer Nicolas de Lacaille (1713–62), who originally classified it as a nebula when he divided Argo Navis into three constellations.

RICH STAR FAMILY
On the edge of NGC 2477 sits the magnitude 4.5 star, b Puppis (shown to the right of the image) which adds to the enjoyment of observing this field.

CANIS MAJOR

Canis Major, the Great Dog, is famous as the constellation that contains the brightest star of the entire night sky—Sirius—a brilliant white star with a magnitude of -1.4.

The Milky Way band flows through this prominent constellation, but as the view in this direction is almost straight out of the plane of our galaxy, it does not offer a spectacularly dense starfield. Nonetheless, Canis Major is a prominent and well-known constellation, representing the larger of the two hunting dogs belonging to Orion, the Hunter. As the Earth rotates, the two dogs seem to follow the Hunter across the night sky. Canis Major is home to six bright stars, including the brilliant Sirius (Alpha Canis Majoris), that are arranged in an easily recognized pattern. Sirius is almost impossible to overlook when stargazing, and the constellation may therefore quickly be found once you have located the Dog Star. Nevertheless, there are other pointers to Canis Major, and they do not come better than Orion's Belt, a straightish line of three stars in the constellation of Orion (see pp.134–135). If you draw an imaginary line through the Belt's three stars, and extend it out from Orion's right side, the next stop is Sirius. Canis Major is also the location of the open star clusters NGC 2362—the Tau (τ) Canis Majoris Cluster—and M41.

M41 👁

An open star cluster lying around 2,300 light-years away, M41 contains approximately 80 stars of 7th magnitude and fainter. It is just visible to the naked eye, and was known even to the ancient Greeks. Around 25 light-years in width, it can be found almost exactly four degrees south of Sirius. M41 appears in the night sky as a hazy patch whose stars are scattered over an area of sky roughly the equivalent of the full Moon's apparent size. Its stars are revealed with binoculars, while a telescope will show chains of stars radiating from the cluster's center. M41 sits on the zenith of the celestial sphere, and it is most visible in the southern hemisphere during January.

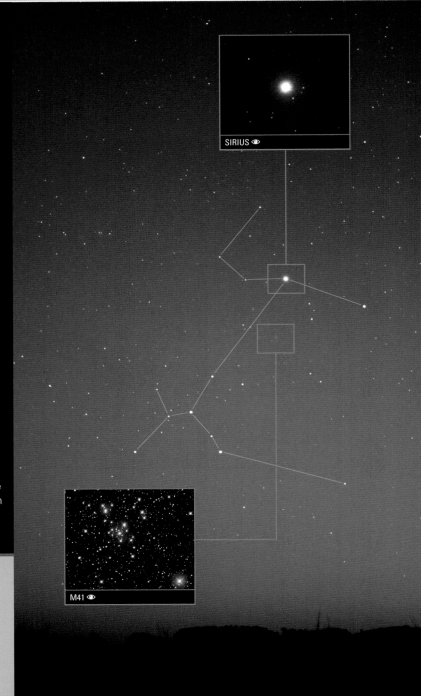

SIRIUS 👁

M41 👁

WIDE VIEW 👁
Canis Major, the larger of Orion's two "dogs," is one of the many bright constellations that light up the summer skies in the southern hemisphere.

CANIS MAJOR (CMa)

WIDTH / DEPTH	🖐/🖐🖐
SIZE RANKING	**43rd**
FULLY VISIBLE	**56°N–90°S**

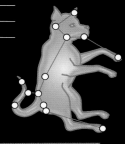

SKY MAP
Canis Major is the 43rd-largest constellation. The Great Dog stands on its hind legs in the sky and holds the star Sirius in its jaws.

OBSERVING CANIS MAJOR

MAJOR STARS	MAGNITUDE	FEATURES
Alpha (α) Canis Majoris *Sirius*	-1.4	The brightest star in the night sky, known as the Dog Star. It has a white dwarf companion, Sirius B.

NOTABLE OBJECTS	MAGNITUDE	FEATURES
M41	4.5	An open star cluster 2,300 light-years away that appears about the size of the full Moon.
NGC 2360	7.2	A dense, wedge-shaped open star cluster best in a small to medium telescope.
NGC 2362	4.4	An open star cluster of 40 stars also known as the Tau (τ) Canis Majoris Cluster.

NGC 2362 👁
This open star cluster, here colored red in an infrared image, is easy to find with the naked eye, as its main star is the magnitude 4.4 Tau (τ) Canis Majoris. NGC 2362 is around 25 million years old and appears rounded through a telescope. The haze that surrounds it is a remnant of the nebula that created it.

Sirius, also known as the **Dog Star**, lies just **8.6** light-years away from the **Solar System**.

Sirius 👁
It is no surprise that such a bright star has been noted throughout history. In ancient Egypt, the first appearance at dawn of Sirius, known as Sothis, occurred before the annual flooding of the Nile, so it was used as an important calendar device. In 1862, several years after it was noted that Sirius had a slight "wobble" in its position, a small companion star was discovered. It orbits Sirius every 50 years and is the cause of the wobble. However, the star itself turned out to be more significant, as it was the first white dwarf to be found.

DOG STAR
Sirius is twice as large as the Sun, and 23 times as luminous. This artist's impression shows how its blue-white companion, to the far right, might appear.

CETUS AND NORMA

The faint constellations of Cetus, the Sea Monster, and Norma, the Set Square, each contain objects that are faint but still impressive. Cetus is home to the star Mira, and Norma has the stunning planetary nebula Shapley 1.

Cetus straddles the celestial equator south of Pisces and Aries, and is stretched over a rather large, empty part of the sky. Only a couple of bright stars highlight this extensive but faint constellation. Cetus is most famous for a long-period variable star that pulsates with a vast change in brightness over a long period of time. The star in question is Omicron (o) Ceti, or Mira. This was discovered by the German astronomer David Fabricius in 1596. It became the prototype of a new class of variable star; the term "Mira-type" was designated to all subsequently discovered variables that acted in the same way. The small constellation Norma was introduced in the 1750s by Nicolas de Lacaille, and resembles a draftsman's set square. It lies in the Milky Way near Scorpius, and is even more inconspicuous than Cetus. Its brightest star is Gamma (γ) Normae, with a magnitude of only 4.1. However, Norma does boast some fine star clusters in the form of NGC 6067 and NGC 6087.

Mira 👁 📷
Omicron (o) Ceti, also known as Mira, is a notable example of a pulsating variable star (see p.19). Some variables are not single stars, but rather a binary pair in mutual orbit—the light fluctuates as one star moves in front of the other, when seen from our point of view. Other variable stars pulsate due to internal forces caused by age, and Mira, translated as "the Wonderful," is one of these. Mira is 300 times larger than the Sun and 15,000 times as bright. Observed from Earth, it varies enormously in luminosity during its 332-day period, during which time it ranges between 2nd and 9th magnitude—meaning it is often invisible to the naked eye. A long-term observer can record this change in action.

CETUS

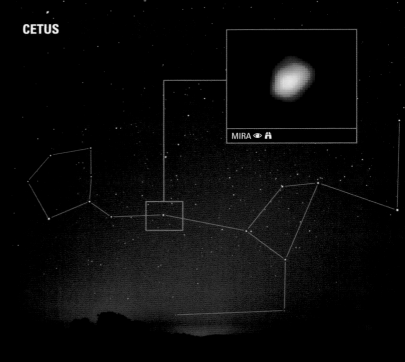

WIDE VIEW 👁
Cetus is the fourth-largest constellation. However, it lies in a large area of the sky's equatorial region that is almost devoid of any bright stars.

NORMA

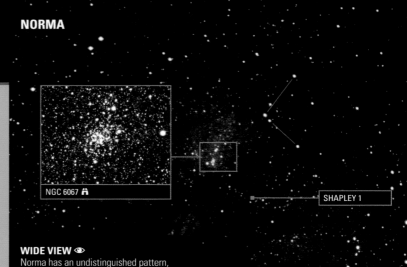

WIDE VIEW 👁
Norma has an undistinguished pattern, consisting of four faint stars roughly arranged in a right-angle, but it offers wonderful views of the Milky Way band.

CETUS (Cet)

WIDTH / DEPTH	
SIZE RANKING	**4th**
FULLY VISIBLE	**65°N–79°S**

SKY MAP: CETUS

Cetus is just below the line of the ecliptic, which means the Solar System's planets will occasionally pass through the constellation.

OBSERVING CETUS

MAJOR STARS	MAGNITUDE	FEATURES OF INTEREST
Alpha (α) Ceti *Menkar*	2.5	A red optical double star with the mag. 5.6 blue companion 93 Ceti.
Omicron (o) Ceti *Mira*	2.0–10.0	A long-period variable, taking 332 days to vary in brightness.

NOTABLE OBJECTS	MAGNITUDE	FEATURES OF INTEREST
M77	8.9	A good telescope is needed to see this spiral galaxy, 47 million light-years away.

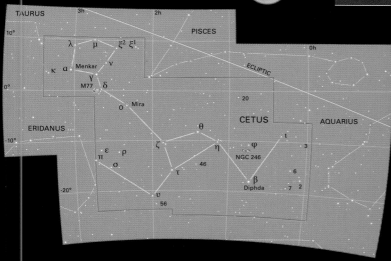

M77 ✸
One of the largest galaxies in the Messier catalog, M77 is a Seyfert-type galaxy with an extremely bright center. It resembles a fuzzy star through smaller telescopes.

A **sea monster** in ancient **Greek** legend, Cetus is also known as the **Whale**.

NORMA (Nor)

WIDTH / DEPTH	
SIZE RANKING	**74th**
FULLY VISIBLE	**29°N–90°S**

SKY MAP: NORMA

Norma was created from stars of the neighboring constellations Scorpius, Ara, and Lupus, and none of its stars have names.

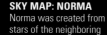

OBSERVING NORMA

MAJOR STARS	MAGNITUDE	FEATURES OF INTEREST
Gamma-2 (γ²) Normae	4.1	A visual double star with the mag. 5.0 star Gamma-1 (γ¹) Normae.

NOTABLE OBJECTS	MAGNITUDE	FEATURES OF INTEREST
NGC 6067	7.7	An open star cluster with over 100 stars, lying around 4,600 light-years away.
NGC 6087	5.4	A bright uneven-looking open star cluster suitable for binoculars.
Shapley 1	13.0	A great symmetrical planetary nebula for larger telescopes.

SHAPLEY 1 ⚖

This planetary nebula is a tremendous object for the more dedicated stargazer with a large telescope. It is perfectly angled for viewing, with a virtually circular hollow "ring" of material lying head-on to the Earth. A star of magnitude 14 sits within the ring's center.

AQUARIUS

Aquarius, the Water Bearer, contains some great deep-sky objects, including the closest planetary nebula to Earth. It is also home to the Delta (δ) Aquariids meteor shower that occurs annually.

Aquarius is an old constellation—known even to the ancient Babylonians—partly due to its positioning along the ecliptic, the yearly path that the Sun appears to make around the Earth in the sky. The Sun passes through Aquarius from February 16 to March 11. As all the objects in the Solar System also move on or close to this line, Aquarius—or in fact any of the zodiac's 12 constellations—may be visited from time to time by the planets of the Solar System. These can make a welcome addition to Aquarius's otherwise faint and relatively ordinary appearance. For example, its brightest star, Alpha (α) Aquarii, or Sadalmelik, shines at a magnitude of only 2.9. Aquarius's stars do have some pattern— particularly the easily recognized asterism of the Water Jar, a Y-shape formed by the stars Gamma (γ), Zeta (ζ), Eta (η), and Pi (π) Aquarii. However, Aquarius is probably best found by observing the Square of Pegasus or the Circlet in Pisces, a short distance away, and tracing a line to the constellation from there. It can also be located using the Summer Triangle (see pp.108–109).

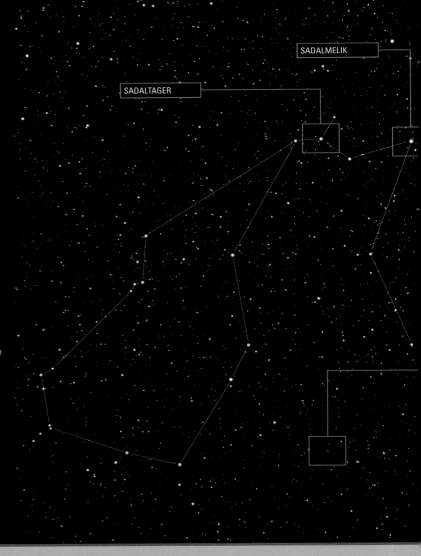

SADALMELIK

SADALTAGER

NGC 7009, the Saturn Nebula 🏃
The Saturn Nebula is so named because when viewed through a large telescope, it bears a striking resemblance to the planet Saturn—including appendages that bring to mind the latter's rings. Smaller telescopes, however, show it simply as a greenish disc. NGC 7009 is a fine example of a planetary nebula. It was originally a low-mass star that eventually transformed into a white dwarf star of magnitude 11.5. The nebula was shaped by material, ejected early in its creation, that confined the subsequent stellar winds into jets. Comparing NGC 7009 to NGC 7293, the Helix Nebula (see opposite), shows what a variety of shapes these celestial objects can take.

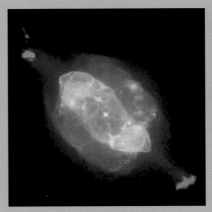

RINGED NEBULA
The ringlike extensions that give the Saturn Nebula its name can be discerned in this Hubble false-color image (left). In a smaller telescope, however, the nebula appears as a blue-green disc (above).

AQUARIUS (Aqr)

WIDTH / DEPTH	
SIZE RANKING	**10th**
FULLY VISIBLE	**65°N–86°S**

SKY MAP

Aquarius sprawls over 980 square degrees of the night sky, making it the 10th-largest constellation. To the Greeks, it represented Ganymede, the gods' cup-bearer.

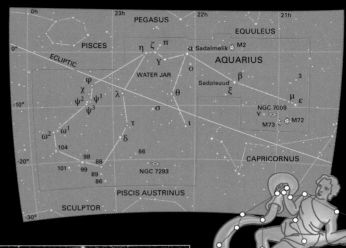

FUZZY CLUSTER

One of the largest-known globular clusters, M2 was discovered in 1746 by the Italian astronomer Jean-Domenique Maraldi. It lies near Aquarius's border with Equuleus and is too faint to see with the naked eye. However, it is fairly easy to find with binoculars or a small telescope, appearing as a fuzzy star.

NGC 7293, THE HELIX NEBULA

WIDE VIEW

Aquarius can be overlooked, as it lies close to the Milky Way and the bright constellations of Lyra, Aquila, and Cygnus. Its distinctive Water Jar is center, top.

NGC 7293, the Helix Nebula

With a small telescope and dark skies, the Helix Nebula—which due to its ocular appearance is informally known as the "Eye of God"—may just be seen as a pale fuzzy disc. Larger instruments are needed to reveal more of its features, and its colors are only revealed when techniques such as photography are used. The Helix Nebula has the largest apparent visual size of any planetary nebula in the skies—it is more than half the width of the full Moon—and it is reckoned to be the closest planetary nebula to Earth, at around 650 light-years' distance. It is about 2.2 light-years across, just over half the distance from our Sun to its nearest neighbor star, Proxima Centauri in Centaurus.

OBSERVING AQUARIUS

MAJOR STARS	MAGNITUDE	FEATURES
Alpha (α) Aquarii *Sadalmelik*	3.0	The second-brightest star in the constellation, after Beta (β) Aquarii, Sadalsuud, at mag. 2.9.
Zeta-2 (ζ²) Aquarii *Sadaltager*	4.4	A double star with Zeta-1 (ζ¹) Aquarii as its mag. 4.6 companion, 5" away.

NOTABLE OBJECTS	MAGNITUDE	FEATURES
M2	6.5	A globular cluster: an island of stars 55,000 light-years distant that appears as a fuzzy blob in binoculars.
NGC 7009, the Saturn Nebula	8.0	A planetary nebula that resembles the appearance of Saturn through a telescope.
NGC 7293, the Helix Nebula	7.3	The largest visual planetary nebula, almost half the apparent width of the full Moon.

CAPRICORNUS AND SCUTUM

These constellations are difficult to find, as their stars are faint. However, they reward patience: Capricornus is home to a fine globular cluster, while Scutum lies on the bright Milky Way band.

Sitting in the same area of the sky, Scutum, the Shield, and Capricornus, the Sea Goat, share a similarity: their brightest stars are only 3rd- and 4th-magnitude, meaning neither particularly stands out in the night sky. These are nevertheless contrasting constellations. Capricornus is the smallest constellation of the zodiac, and is said to represent the Babylonian water god Ea, whose symbols were a fish and a goat. It is located in a fairly featureless part of the sky. However, Scutum, to the north of Capricornus, lies in a wonderful meandering section of the Milky Way. In fact, the cloudy area here is the brightest part of the Milky Way outside that in Sagittarius, the Archer. Using Sagittarius is the easiest way of finding Scutum, as this small constellation lies only a short way north of the Archer's "Teapot" asterism. To find Capricornus, extend an imaginary line from Vega, in Lyra, out through Altair, in Aquila, and you will find the two main stars of the Sea Goat: Alpha (α) Capricorni (Algedi) and Beta (β) Capricorni (Dabih).

M11, the Wild Duck Cluster

From a dark location, the Wild Duck Cluster—lying 5,600 light-years from Earth, in Scutum—can be seen with the naked eye. It appears as a brighter misty patch of the Milky Way, about half the apparent size of the full Moon. Binoculars are a step forward, but they will still show this star family as a smudge of light. Only a telescope will distinguish some of the 3,000 or so stars that make up this triangular-shaped group, which appears to replicate the flight-pattern of a flock of birds. M11 is located in an area of the constellation known as the Scutum Star Cloud, one of the brightest parts of the Milky Way. Near the V-shape's apex is Scutum's brightest star, Alpha (α) Scuti, shining at magnitude 3.9.

CAPRICORNUS

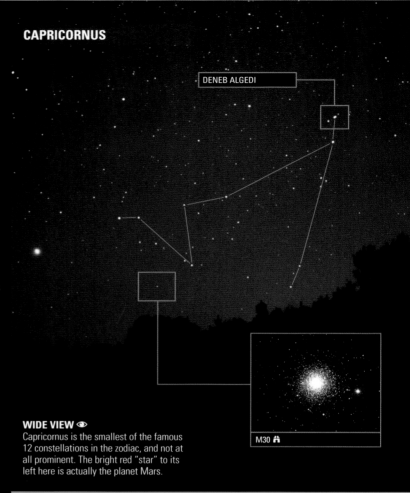

DENEB ALGEDI

M30

WIDE VIEW 👁
Capricornus is the smallest of the famous 12 constellations in the zodiac, and not at all prominent. The bright red "star" to its left here is actually the planet Mars.

SCUTUM

M11, THE WILD DUCK CLUSTER

WIDE VIEW 👁
The fifth-smallest constellation, Scutum lies between Aquila and Sagittarius, and can be difficult to locate if dark skies also show the bright Milky Way.

CAPRICORNUS (Cap)

WIDTH / DEPTH	🖐🖐 / ✊✊
SIZE RANKING	**40th**
FULLY VISIBLE	**62°N–90°S**

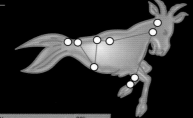

SKY MAP: CAPRICORNUS
The brightest star of this roughly triangular constellation is Delta (δ) Capricornus, Deneb Algedi, at the Sea Goat's tail.

OBSERVING CAPRICORNUS

MAJOR STARS	MAGNITUDE	FEATURES OF INTEREST
Alpha-2 (α²) Capricorni *Secunda Algedi*	3.6	This forms an optical double with the mag. 4.3 Alpha-1 (α¹) Capricorni, Prima Algedi. Both stars are golden in color and are separated by 61' 30".
Beta (β) Capricorni *Dabih*	3.1	A double star when viewed in binoculars, with a mag. 6.1 companion 3' 25" away. Its stars are yellow and white.

NOTABLE OBJECTS	MAGNITUDE	FEATURES OF INTEREST
M30	5.8	An open star cluster that lies around 6,000 light-years away from Earth and is a fine sight through a telescope.

HAZY PATCH ↗
A globular cluster with a collapsed core, M30 is a 90 light-years-wide ball of stars, moving toward the Solar System at 113 miles (182km) per second. It is visible as a hazy patch through a small telescope.

SCUTUM (Sct)

WIDTH / DEPTH	✊ / ✊
SIZE RANKING	**84th**
FULLY VISIBLE	**74°N–90°S**

SKY MAP: SCUTUM
The original name of this constellation was Scutum Sobiescianum, and the leading star, Sobieski, references this.

astronomer **Hevelius** named
...tum in **honor** of his patron
...Sobieski, the king of **Poland**

OBSERVING SCUTUM

NOTABLE OBJECTS	MAGNITUDE	FEATURES OF INTEREST
M11, the Wild Duck Cluster	5.8	An open star cluster that sits 5,600 light-years away from the Earth and is a fine telescopic object.
M26	8.0	An open star cluster, not as visually fine as M11, but still worth locating in a telescope.

DUCKS IN FLIGHT
Considered to be one of the finest open star clusters in the sky, M11's V-shaped, "duck-flight" pattern is less apparent in photographs.

SAGITTARIUS

Sagittarius, the Archer, is dominated by a part of the Milky Way that is luminous, meandering, and full of bright nebulae and rich star clusters.

On the border of Sagittarius with Scorpius lies a point that marks the location of our galaxy's center, so when we look toward this prominent constellation we are gazing into the heart of the Milky Way. It is not surprising, therefore, that Sagittarius is filled with dark and bright nebulae, globular clusters, and wonderful open star clusters, all set within a rich starfield. The Milky Way's brightness here can sometimes quite overwhelm the star pattern. However, the main luminaries of Sagittarius, even though they sit within the galactic band, form a fairly bright and recognizable asterism known affectionately as the Teapot. Strangely, the stars Alpha (α) and Beta (β) Sagittarii—Rukbat and Arkab respectively—are not located near this group, nor are they the constellation's brightest; they are two faint stars that form the Archer's right foreleg.

A supermassive **black hole** may lie in **Sagittarius A***, the **center** of the Milky Way.

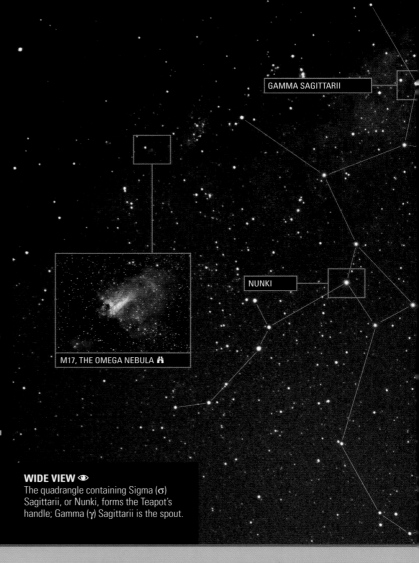

GAMMA SAGITTARII

NUNKI

M17, THE OMEGA NEBULA

WIDE VIEW 👁
The quadrangle containing Sigma (σ) Sagittarii, or Nunki, forms the Teapot's handle; Gamma (γ) Sagittarii is the spout.

M8, the Lagoon Nebula
This nebula is one of the largest in the sky. It is lit by the radiation of stars that are being created within it—its bright clouds are a good example of an emission nebula. M8 appears through binoculars as a grayish oval with an off-center brighter patch. Small, darker features, known as Bok globules, can also be viewed with a telescope. These condensed areas of the nebula are protostellar clouds, caused by gravity pulling in the surrounding dust and gas. A single star, or a multiple star system, can eventually form in these globules as a result of nuclear reactions caused by rising temperatures in their centers.

DARK DUST LANE
Larger telescopes will reveal the dark "lagoon" of dust meandering through the nebula's cloud, giving M8 its common name.

SAGITTARIUS (Sgr)

WIDTH / DEPTH	
SIZE RANKING	**15th**
FULLY VISIBLE	**44°N–90°S**

SKY MAP
The northern part above the Teapot is the main area to search for this constellation's amazing deep-sky objects.

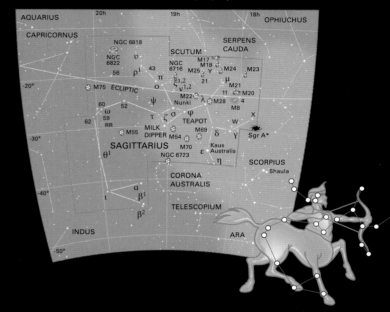

OBSERVING SAGITTARIUS

MAJOR STARS	MAGNITUDE	FEATURES
Alpha (α) Sagittarii *Rukbat*	4.0	A bluish star 170 light-years from Earth.
Beta-1 (β¹) Sagittarii *Arkab*	4.0	A blue double star for naked-eye viewing, with the white, mag. 4.3 companion Beta-2 (β²), Arkab Posterior.
Epsilon (ε) Sagittarii	1.8	The constellation's brightest star.

NOTABLE OBJECTS	MAGNITUDE	FEATURES
M8, the Lagoon Nebula	5.8	A 5,200-light-year-distant emission nebula with a cluster of around 25 stars within.
M17, the Omega Nebula	7.0	In binoculars this emission nebula resembles a triangular slice of gray cake.
M20, the Trifid Nebula	9.0	A small, roundish emission nebula that is best seen with a telescope.
M22	5.1	An impressive globular cluster just smaller than the apparent size of the full Moon.
M23	5.5	An open cluster 2,150 light-years from Earth.
M24, the Sagittarius Star Cloud	4.5	A starfield covering a patch of the Milky Way that is three times the apparent diameter of the full Moon.
M25	4.6	An open star cluster, containing around 30 stars, that appears as a smudge to the naked eye.

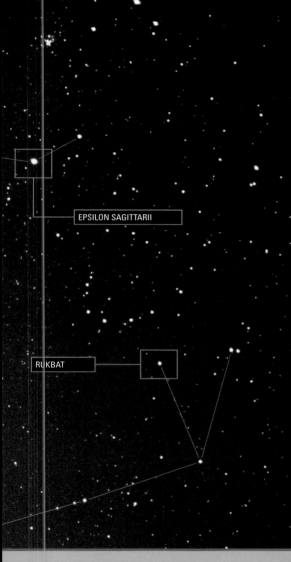

EPSILON SAGITTARII

RUKBAT

M17, the Omega Nebula

This nebula is a bright cloud of glowing gas with a distinctive curved appearance that can be likened, as the name suggests, to the Greek capital letter Omega (Ω), although its alternative names of the Swan or Horseshoe Nebula also give some indication of its shape. Like all emission nebulae, M17 appears slightly pale-green to the naked eye, but photography brings out the red coloring, which is caused by ionized hydrogen. M17 can be glimpsed through binoculars, but its true shape is better discerned through a small telescope. The latter will also reveal a small star cluster, NGC 6618, within the cloud. Both the nebula and the star cluster lie about 4,900 light-years from Earth.

SCORPIUS

Scorpius, the Scorpion, has a number of attractions for stargazers: the Milky Way, deep-sky objects, and bright stars in one of the most recognizable constellation patterns.

It is difficult for observers in the northern hemisphere to truly appreciate the wonders of Scorpius, for it lies well into the southern skies. Its main stars are led by Antares, or Alpha (α) Scorpii, and their arrangement south needs little imagination to be interpreted as a scorpion with a curled tail ready to sting. The creature has lost its claws, however, as the Romans "removed" them in order to create the constellation of Libra, the Scales, which borders Scorpius. Antares itself is a supergiant that fluctuates in brightness from magnitude 0.9 to 1.2 every four to five years. It has an enormous diameter, around 800 times that of the Sun—if placed in the Solar System, this star's surface would extend beyond the orbit of Mars; fitting, perhaps, as like the planet it is colored a bright red. Indeed, this color gives the star its name: Antares roughly translates as "anti-Ares" or "rival of Mars." Ares was the Greek name of the Roman god of war, Mars. Another of Scorpius's noteworthy stars is Delta (δ) Scorpii. It is normally of magnitude 2.3, but since 2000 it has unexpectedly been variable and brighter by up to about 0.6 magnitude.

NGC 6231, the Table of Scorpius 👁

Perhaps only just over 3 million years old, NGC 6231 is a prominent, very young open star cluster of around 100 stars, located just too far south in the sky to be included in Charles Messier's original celestial catalog. The Table of Scorpius sits 5,900 light-years away from the Solar System and, to the naked eye, it seems to be merely a tightly knit group of stars, but its individual stars offer a truly marvellous visual display when viewed in binoculars or a small telescope. Just outside the main group is the 5th-magnitude star Zeta-1 (ζ^1) Scorpii, which may actually also be a member of this star family. This star has a 4th-magnitude optical companion that lies much closer to the Solar System.

M6, THE BUTTERFLY CLUSTER 👁

NGC 6231, THE TABLE OF SCORPIUS 👁

WIDE VIEW 👁
The curving-tail design of the Scorpion's stars is without a doubt its main feature; the Scorpion's claws were used to form the nearby constellation of Libra.

SCORPIUS (Sco)

WIDTH / DEPTH	
SIZE RANKING	**33rd**
FULLY VISIBLE	**44°N–90°S**

SKY MAP
The center of the galaxy as seen in the night sky lies just outside Scorpius, thus the wealth and variety of star clusters.

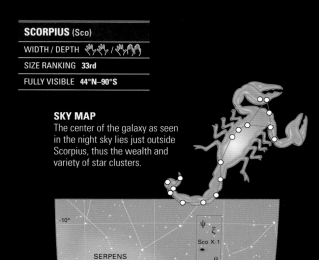

OBSERVING SCORPIUS

MAJOR STARS	MAGNITUDE	FEATURES
Alpha (α) Scorpii *Antares*	1.0	The 16th-brightest star in the sky, 10,000 times brighter than the Sun.
Beta-1 (β¹) Scorpii *Graffias*	2.6	A double star with the mag. 4.9 Beta-2 (β²) Scorpii.
Zeta-1 (ζ¹) Scorpii	4.7	A double star with the brighter mag. 3.6 companion Zeta-1 (ζ¹) Scorpii. These are orange and bluish stars sitting 6' 30" apart.
Mu-1 (η¹) Scorpii	3.0	A naked-eye double with the mag. 3.6 star Mu-2 (η²) Scorpii. The stars are both bluish in color.

NOTABLE OBJECTS	MAGNITUDE	FEATURES
M4	6.0	One of the closest globular clusters to the Solar System at 6,800 light-years' distance from Earth.
M6, the Butterfly Cluster	4.2	An open star cluster sitting just to the north of M7 (below), in a slightly darker part of the Milky Way—making it easier to observe.
M7, the Ptolemy Cluster	3.3	An open star cluster that is a great sight in binoculars, with about 80 stars of mag. 10 or brighter.
M80	7.3	A smallish globular cluster that can be found with perseverance in binoculars. Its center has a very compact appearance.
NGC 6231, the Table of Scorpius	2.5	An open star cluster that looks great through binoculars, as it sits within the band of the Milky Way.

Scorpius is only **fully visible** from Earth locations **south of 40°N**, the latitude of Madrid.

M7, the Ptolemy Cluster 👁

An exquisite open cluster, the starry family M7 is commonly referred to as the Ptolemy Cluster, after the Roman/Greek academic who first described it in 130 CE. M7 is considered one of the best deep-sky objects in Scorpius due to its brightness and size; it is more than twice the apparent size of the full Moon. To the eye, the Ptolemy Cluster looks like a brighter clump of the dense Milky Way starfields, but with binoculars its fine stars are revealed. It lies near the tail of Scorpius, appearing close to a second cluster, M6—the Butterfly Cluster—which is twice as distant as M7 and appears much smaller and fainter.

BRIGHT CLUSTER
M7 is situated in front of the bright Milky Way band, further enhancing its 6th-magnitude stars. It lies around 780 light-years away from Earth.

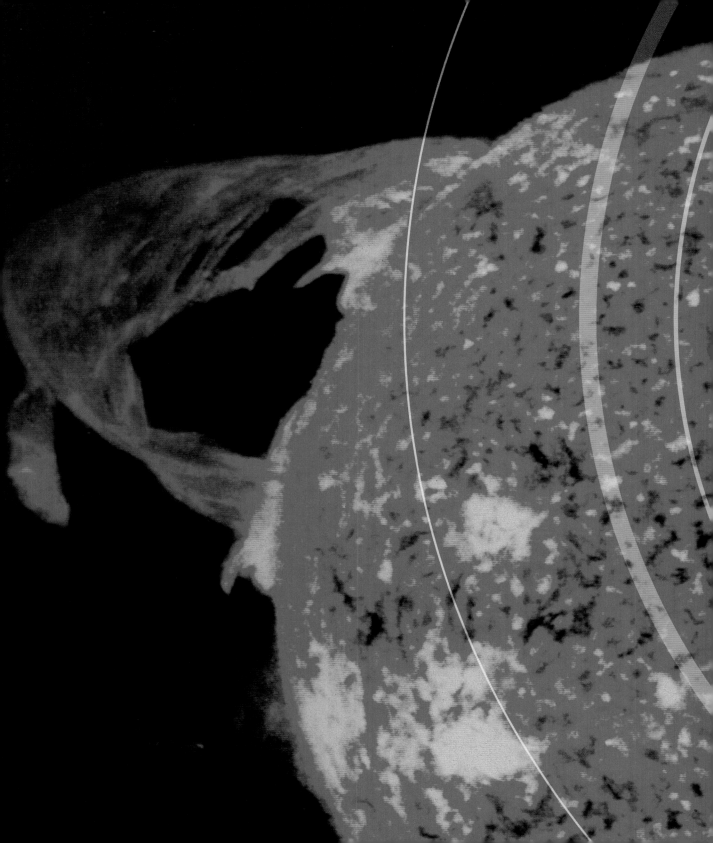

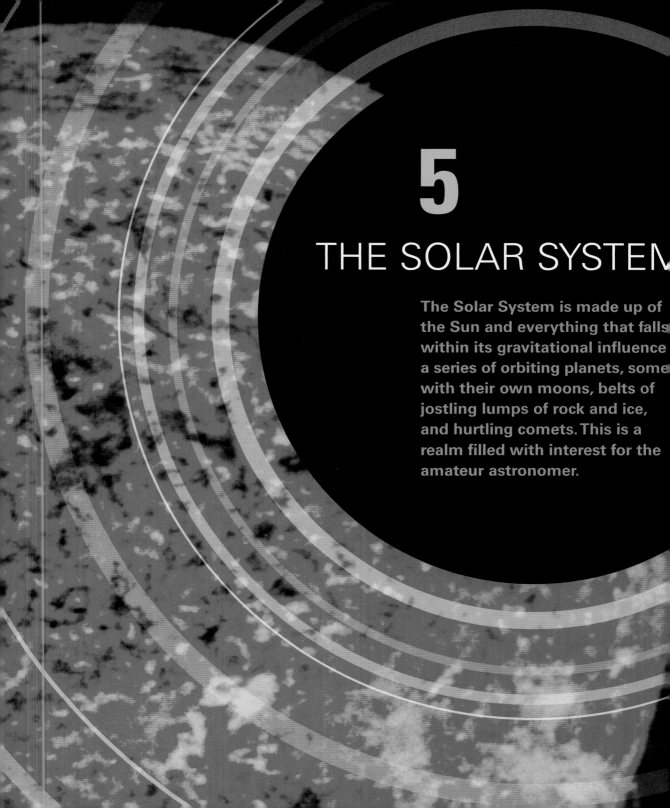

5

THE SOLAR SYSTEM

The Solar System is made up of the Sun and everything that falls within its gravitational influence a series of orbiting planets, some with their own moons, belts of jostling lumps of rock and ice, and hurtling comets. This is a realm filled with interest for the amateur astronomer.

OUR VIEW OF THE SOLAR SYSTEM

While the Universe is worthy of a lifetime's observation, there is a plethora of objects in our own Solar System that make fascinating targets.

Planets in orbit

Without a doubt the three most fascinating planets to observe are Mars, Jupiter, and Saturn; the first due to its proximity to us, the second because of its moons and cloud bands, and the last, of course, for its wonderful ring system. Lying closer to the Sun than does Earth, Venus shines out as a beautiful "star," but little detail can be observed because its surface is always hidden by highly reflective opaque cloud. Beyond Saturn lie Uranus and Neptune. They may be huge planets, but they receive so little of the Sun's light, and are also so far from us, that they appear only as faint objects.

THE MAIN PLAYERS
A total of eight planets are held within the vast gravitational pull of the Sun. Earth is just one of the four rocky planets whose orbits lie closer to the Sun than the Asteroid Belt. Beyond the Belt lies Jupiter, the nearest of four "gas giants."

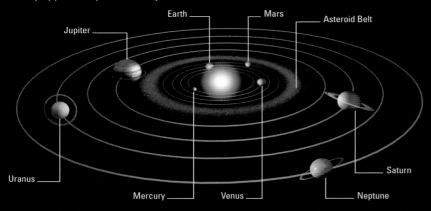

Earth — Mars — Asteroid Belt
Jupiter —
Uranus —
Mercury — Venus — Neptune
Saturn

Perpetual motion and change

Everything we see within the Solar System is moving, from the stately progress of a planet to the momentary flash of a meteor, or "shooting star," and this is what makes our neighborhood in space so fascinating to observe. The planets and our own Moon follow a set path (see facing page), but there are other objects that are not so predictable: comets, for example, that happen to fly in past the Earth may provide a spectacular, if transient, display, even to the naked eye. Atmospheric phenomena such as aurorae can transform the night sky, if only for a few hours, and then there are the winking "eyes" and brilliant flares that indicate the presence of man-made satellites—our own contribution to this space ballet.

"FIREBALL" METEOR
The brightest meteors of all are known as fireballs, and may be seen at any time and location. Some are so bright as to be visible in daylight.

CONJUNCTION
Venus (the brightest), Mars, and Jupiter (far left) are shown here in conjunction. The Moon often joins planets in these alignments, for it also follows the ecliptic.

The path of the planets

The way in which the Solar System was formed (see pp.14–15) resulted in all of its planets orbiting in or near the same plane, like marbles sitting on a table. This plane, when drawn across the sky, is known as the ecliptic (see also p.34), and the constellations that it runs through are the famous 12 that form the zodiac. Therefore all the major planets will be found in (or occasionally just outside the borders of) one of the signs of the zodiac. Astronomical almanacs, magazines, and websites give details of planets visible in various constellations month by month; with this information, it is simply a case of observing that constellation and spotting the extra "star." If this "star" changes position over a series of observations, you can be sure you have located the planet.

Occasionally a fortuitous alignment of positions results in two or more (rarely, as many as five) planets to be viewed in the same line of sight—this is known as a conjunction. When this happens it is possible to gain a whole new perspective on skywatching, for when one knows that one must be looking, for example, past Mars to Jupiter, one gets a real sense of looking out into space, rather than regarding the night sky as our celestial "ceiling."

When paths cross

With most Solar System objects orbiting close to the same plane, there are times when we see two of them come into perfect alignment, one moving in front of the other. When the Sun, Earth, and Moon are involved this leads to a solar or lunar eclipse. If the Sun, Earth, and Mercury or Venus are lined up, then we may see a transit—a time when Mercury or Venus moves across the face of the Sun. Other alignments occur when the Moon moves in front of a planet; a large object obscuring a smaller one in this way is known as an occultation.

OCCULTATION
The Moon, here in crescent phase, is about to cross the path of Jupiter, hiding it from view until the planet "emerges" on the other side.

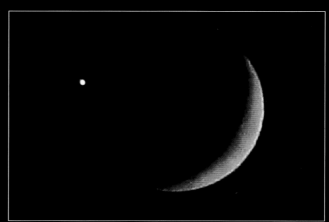

OBSERVING THE PLANETS

When and where we see a planet in the sky, and how it seems to move, result from its position and motion in space relative to Earth.

Relative positions

Mercury and Venus together are known as the inferior planets, which just means that they orbit closer to the Sun than does the Earth. The best time to view them is when they are close to maximum elongation—this is simply the furthest distance that the planet gets from the Sun as seen from our viewpoint (see panel, right). This is especially important if you want to see Mercury, as its smaller orbit means that it is often very close to the Sun. Inferior planets seem to "follow" the Sun down as it sets or appear before sunrise.

The planets with orbits beyond that of Earth—Mars, Jupiter, Saturn, and remote Uranus and Neptune—are described as superior, and these can be visible all night long and at any elevation from the horizon. A superior planet is at its brightest and largest when its orbit brings it closest to Earth, a time called opposition. Any planet at opposition will rise at sunset, reach its highest in the sky at midnight and set at sunrise. This pattern of movement represents the main difference in visibility between the superior and inferior planets.

THE INFERIOR PLANETS

Mercury and Venus, the inferior planets, orbit closer to the Sun than does Earth. For part of their orbits they are hidden in the Sun's glare, and when we do see them, they appear to have phases, like the Moon. When on the far side of the Sun, they would appear as a full disc were they not invisible to us. As they move closer, we begin to see their sunlit side. The best time to view either planet is that of greatest elongation: the point at which the planet is at the greatest possible angle from the Sun as viewed from Earth. At this point we see them in a half-phase. As they move nearer they grow larger, but we only see a crescent phase. The point at which they lie directly between the Earth and the Sun (when, of course, they again disappear from view) is known as inferior conjunction, then once they pass Earth, the process is reversed. At greatest eastern elongation (GE) they are visible in the evening sky just after sunset; at greatest western elongation (GW), they appear in the pre-sunrise dawn skies.

THE INFERIOR PLANETS SEEN FROM EARTH

For simplicity the Earth is shown below as stationary. The inferior planet (here Venus) appears at its largest when in its crescent phase; we see the maximum amount of its disc when it is furthest from us, and smallest.

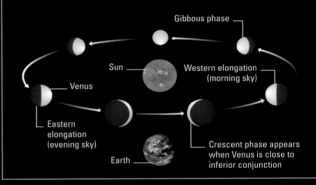

Gibbous phase

Sun

Venus

Western elongation (morning sky)

Eastern elongation (evening sky)

Earth

Crescent phase appears when Venus is close to inferior conjunction

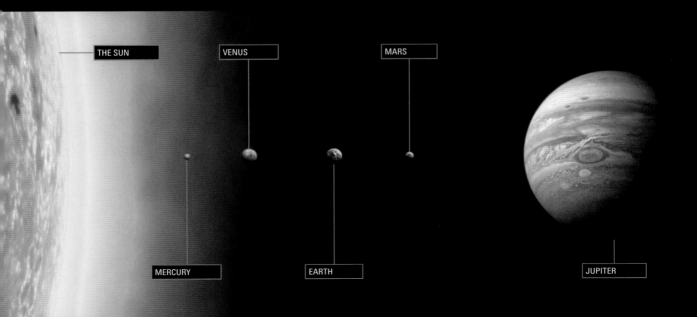

THE SUN

VENUS

MARS

MERCURY

EARTH

JUPITER

What can we see?

All of the planets are visible to the naked eye except Neptune, for which binoculars are required. To the eye they appear as bright or faint stars, but in binoculars, Venus, Mars, Jupiter, and Saturn reveal themselves to be planets by their disc-like shape. In powerful binoculars, the phases of Venus can be discerned. Mercury, Uranus, and Neptune, however, still present themselves as star-like points. A small telescope does not dramatically improve the view of Mercury and Venus, but fascinating detail becomes apparent when observing Mars, Jupiter, and Saturn. Surface features such as the polar caps of Mars and the belts and zones of Jupiter come into view, and Saturn's rings are clearly visible. Through a telescope, Uranus and Neptune finally resolve themselves into faint, blue-green discs.

ENHANCING THE VIEW
The planet Saturn seen, from top left moving clockwise, with the naked eye, in binoculars, through a small telescope, and using CCD imaging.

Retrograde motion

Superior planets are normally observed over time as moving in a west-to-east direction. However, as any one approaches opposition its eastwards movement halts and it moves backward in the sky for a month or so before halting again, then recommencing its normal motion. This is known as retrograde motion. This odd westward shift is due to the difference in orbital speeds between Earth and the planet and it is akin to two cars on a race track: the fast "Earth car" on the inside line passing a slower "planet car" on the outside lane—the moment of passing being opposition, when the two cars are at their closest. If you imagine yourself in a fast car looking out of your side window as you pass a slower one, it would appear to go backward until you had gained enough distance for it to be visible in the rear-view mirror, when it would once again appear to be traveling forward.

GOING INTO REVERSE
A superior planet reaches opposition—its closest to Earth—when the Earth lies directly between it and the Sun, and it is during this time, as we overtake on our faster orbit, that we observe retrograde motion.

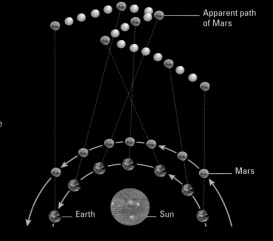

Apparent path of Mars

Mars

Earth

Sun

SATURN

NEPTUNE

CHAIN OF PLANETS
The planets are here arranged in order of their distance from the Sun. Their relative

MERCURY AND VENUS

These two inner worlds could not be more different in appearance: while small and faint Mercury is often called the elusive planet, Venus can be a truly brilliant "star."

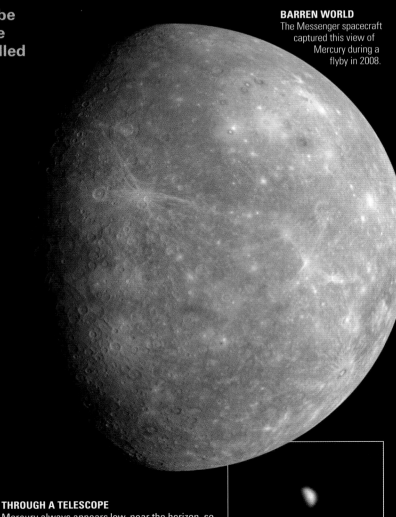

The Sun's neighbors

Mercury is a tiny, rocky world, its surface desolate and heavily cratered by millions of years of bombardment by meteoroids. Venus, further away from the Sun, is a much larger planet. Its surface, shrouded in sulfuric acid clouds in a thick carbon dioxide atmosphere, is a scorching and barren landscape dominated by huge volcanoes and vast solidified lava flows. Venus rotates so slowly that its day is longer than its year—in other words, Venus orbits the Sun in a shorter time than it rotates once on its axis. Venus's orbit also brings it closer to us than any other planet at just under 25 million miles (40 million km), about 100 times further from Earth than our Moon. As the inferior planets (see p.176), orbiting closer to the Sun than does Earth, Mercury and Venus are always seen in "phases," like the Moon—with only a half-disc or crescent, for example, lit by the Sun.

MERCURY PROFILE

MEAN DISTANCE FROM SUN	36 million miles (57.9 million km)
ORBITS THE SUN IN	88 days
ROTATES IN	58 days 15 hours 30 minutes
GREATEST MAGNITUDE	-1.9
FARTHEST FROM EARTH	138 million miles (221.9 million km)
NEAREST TO EARTH	48 million miles (77.3 million km)
GREATEST ANGULAR SIZE	12.9 arcseconds
NUMBER OF MOONS	0

SIZE COMPARISON

Mercury

Earth

THROUGH A TELESCOPE
Mercury always appears low, near the horizon, so observations are always made through our dimming and possibly churning atmosphere. As a result, it appears as a fuzzy rather than sharp-edged disc.

Observing Mercury

Mercury's small size and distance from us and its closeness to the Sun makes observation challenging. It may be seen with the naked eye hanging in the dusk or dawn skies, but its fast orbit means that it does not stay visible for long. A good observing window rarely lasts much longer than a week. A telescope will show its phases as it moves toward us. The planet's maximum apparent size as seen from the Earth is just 12.9 arcseconds—and because of that, even with a telescope, it is only possible to see fleeting glimpses of vague patches on its surface.

TRANSITS

Both Mercury and Venus move fast, but in dawn and twilit skies, low on the horizon with no background stars in view, they can be hard to track. There is one time when both planets can be witnessed actually traveling in their orbits, and this is during a transit—the time when Venus or Mercury cross the disc of the Sun as observed from Earth. For Mercury this event happens about 12 or 13 times per century, but with Venus such events are rare and much-anticipated—it has one pair of transits 8 years apart, followed by a 105-year gap, then a further pair 12 years apart, followed by another, great gap of 121 years.

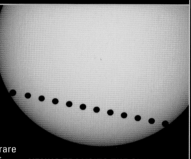

VENUS TRANSITS THE SUN
Transits must only be observed (here in a series of time-lapse photographs) with full safety precautions (see p.54).

VENUS PROFILE

MEAN DISTANCE FROM SUN	67.2 million miles (108.2 million km)
ORBITS THE SUN IN	224.7 days
ROTATES IN	243 days 0 hours 30 minutes
GREATEST MAGNITUDE	-4.6
FARTHEST FROM EARTH	162 million miles (261 million km)
NEAREST TO EARTH	23.7 million miles (38.2 million km)
GREATEST ANGULAR SIZE	66 arcseconds
NUMBER OF MOONS	0

SIZE COMPARISON

Venus

Earth

Observing Venus

To the unaided eye, when anywhere near its brightest, Venus outshines everything bar the Moon. With sunlight always reflecting white off its clouds, together with its closeness to us between the times of eastern and western elongation (see p.176), Venus well deserves its familiar names, the Evening or Morning Star. Being further from the Sun than Mercury, Venus does not travel so fast, and so is visible for much longer periods than its inner neighbor—it can be up in the evening or morning skies for around six months at a time. Venus appears largest to us in its crescent phase, when it is so close to Earth that technically its phase should be apparent to the unaided eye, although a stargazer's vision must be exceptionally good to overcome the dazzling brilliance of the planet. With binoculars its phases are easily visible. The best time to view Venus with a telescope is not in a dark sky, as the planet's brightness is overwhelming, but around sunset as the planet's contrast is lower against a sky that is only just beginning to darken.

THROUGH A TELESCOPE
A telescope will show the sunlit part—the "phase"—of Venus as dazzling white but with possible, very subtle atmospheric features.

MYSTERIOUS WORLD
The Pioneer Venus Orbiter took the first full-disc picture of Venus in 1979.

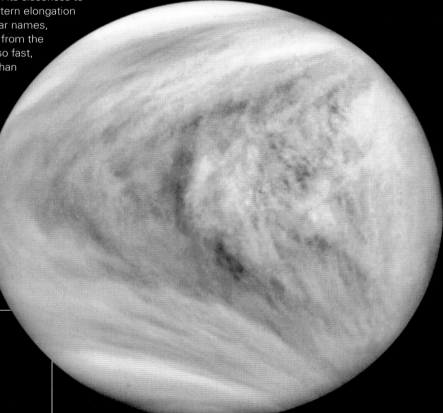

MARS

With its brightness and striking red color, Mars has fascinated observers for thousands of years, and as technology has evolved, it has been subject to intense scrutiny and speculation.

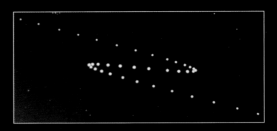

CHANGING DIRECTION
Mars is a good target for observing retrograde motion (see p.177). In this multi-exposure image of its path over several months it appears to travel "backward," then forward again.

The red planet

Even though it is only around half the diameter of our planet, Mars is perhaps the most Earthlike of all the planets in the Solar System. Its surface—the only planetary surface we can observe in any detail from Earth—is relatively barren, marked with volcanoes, craters, and vast chasms, the largest of which (the Valles Marineris) stretches over 2,500 miles (4,000km). The planet's largest volcano, Olympus Mons, is truly gargantuan, measuring around 400 miles (650km) across and reaching 15 miles (24km) through the Martian sky. The surface is covered in dust from the red rock rich in iron oxides, which can be whipped up into fierce storms. Mars has two moons, Phobos and Deimos, probably asteroids captured by the planet's gravitational pull.

Observing Mars

Mars is a very swift planet, traveling through almost one constellation per month across the sky when away from opposition. For much of the time it can sit almost unnoticed in the night sky, looking like an ordinary "star" of magnitude 1.7. However, every two years and two months a dramatic increase in brightness takes place when it comes into opposition; its greater proximity to Earth at this time makes it the second-brightest planet in the sky after Venus. Then, as the Earth's faster orbital speed takes us away from Mars, its size and thus brightness seems to diminish. In fact, Mars has only a two-month window centered around opposition when it is best observed, so planet watchers await these times with great anticipation. Mars's elliptical orbit brings it especially close to Earth at some oppositions; these occur every 15 or 17 years.

MARS PROFILE

MEAN DISTANCE FROM SUN	141.6 million miles (227.9 million km)
ORBITS THE SUN IN	686.9 days
ROTATES IN	24 hours 37 minutes 23 seconds
GREATEST MAGNITUDE	-2.9
FARTHEST FROM EARTH	249.4 million miles (401.3 million km)
NEAREST TO EARTH	33.9 million miles (54.5 million km)
GREATEST ANGULAR SIZE	25.7 arcseconds
NUMBER OF MOONS	2

SIZE COMPARISON

Mars   **Earth**

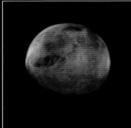

CHANGING SIZE
This composite of two Hubble images shows how the apparent size of Mars varies through a telescope, from when it is at conjunction and farthest away from us (right), with the Sun lying between Earth and Mars, to when it is closest to us, at opposition (left). The two-month window during which this enhanced view is possible provides a fantastic opportunity to observe Mars in detail.

Discovering Mars

Many civilizations have named this blood-red planet after their god of war, from the Babylonian "Nergal" to the Roman "Mars." The increased resolution of telescopes in the 19th century led to a new fascination with the planet, as its surface and atmospheric features were revealed. At the end of the 1800s the most famous illusory markings on Mars, the "canals," were first described by the great Italian astronomer Giovanni Schiaparelli. Even though they do not exist, they formed part of a planet-wide map that he drew, and the names Schiaparelli gave many surface features are still those in use today.

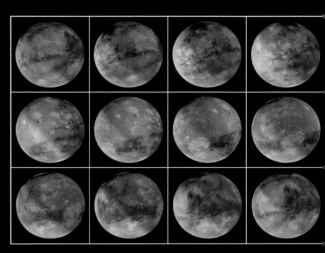

A CHANGING VIEW

This digital illustration shows a complete rotation of Mars. Top row: the massive, scar-like Valles Marineris can be seen just below the equator. Central row: Olympus Mons moves across as the "thumbhole" of a bowling-ball-like arrangement. Bottom row: the orange disc of the Hellas Basin impact scar lies just above the south pole, with the vast, dark Utopia Planitia plain above it.

The Martian landscape

The most prominent features on Mars can be seen as light and dark markings with even a small telescope when Mars is at its closest. The polar caps, formed of water ice and frozen carbon dioxide, are the easiest features to see. With a larger telescope the main topographical features gain definition, and storms and even white clouds may also be discerned. Mars spins at a very sedate pace—its day being only 37 minutes longer than our own—which allows plenty of time to observe and sketch features before they rotate out of view.

THE MOONS OF MARS

Mars has two moons: inner 8-mile (13-km) diameter Phobos, and outer 14-mile (23-km) diameter Deimos. Their small sizes make them faint targets for observers and they are often overwhelmed by the brightness of Mars itself, especially the smaller Phobos (seen below, far right, "transiting" or crossing the Martian landscape). However, when Mars is at opposition and the moons are situated as far from the planet as possible, these tiny worlds—at magnitude 11.7 (Phobos) and 12.8 (Deimos)—can be seen through medium-sized telescopes.

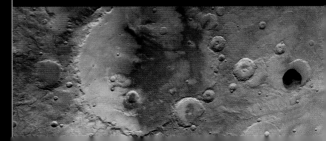

JUPITER

Beyond the orbits of the inner rocky planets lies the realm of the gas giants—the first and largest of which is Jupiter.

A world of storms

Jupiter is the largest planet in the Solar System, at 11 times the diameter of Earth. It is a vast globe banded by powerful storms: its atmosphere of hydrogen and helium is constantly driven by winds of up to 388 mph (625kph). Jupiter's immense gravity has meant it has clung on to or captured a vast number of satellites, or moons, and it also has a faint ring system. While Jupiter is composed of the same materials as the Sun, it is a common misconception to view it as a failed star. The word "failed" assumes it once nearly had enough mass to start some stellar nuclear processes, which it did not. But it is true to say that if Jupiter was made much larger, and contained 12 times as much material, its central temperature would be high enough for it to qualify as a quasi-stellar object known as a brown dwarf.

Observing Jupiter

Jupiter is easy to spot with the naked eye, because even at its faintest it outshines Sirius, the brightest star. Its slow orbit means it travels slowly across the night sky, at a rate of only about one constellation per year. It comes into opposition (when it is closest to Earth and appears brightest) every 13 months. At these times the planet is a great target for astronomers, as even a pair of binoculars will reveal the planet as a disc accompanied by four main moons. One thing that is noticeable in a telescope is the "squashed" shape of the planet, which is caused by its rapid rotation. This throws out the equator to almost 88,850 miles (143,000km) across, while the pole-to-pole diameter is squeezed down to a lesser 83,260 miles (134,000km). Jupiter's fast rotation also makes it a particularly interesting planet to observe, because its surface appearance is constantly changing.

MOON ORBITING JUPITER
The closest of the main moons, Io, here casts a distinct shadow on the planet, despite being some 310,000 miles (500,000km) away.

JUPITER PROFILE

MEAN DISTANCE FROM SUN	483.6 million miles (778.3 million km)
ORBITS THE SUN IN	11.86 years
ROTATES IN	9 hours 50 minutes 30 seconds
GREATEST MAGNITUDE	-2.9
FARTHEST FROM EARTH	601.5 million miles (968.1 million km)
NEAREST TO EARTH	365.7 million miles (588.5 million km)
GREATEST ANGULAR SIZE	59 arcseconds
NUMBER OF MOONS	79

SIZE COMPARISON

Earth

Jupiter

JUPITER THROUGH A TELESCOPE
With a small telescope, you can make out Jupiter's principal moons, which appear as bright stars on either side of the planet. The dark stripes on either side of the planet's equator, known as the north and south equatorial belts, should also be visible.

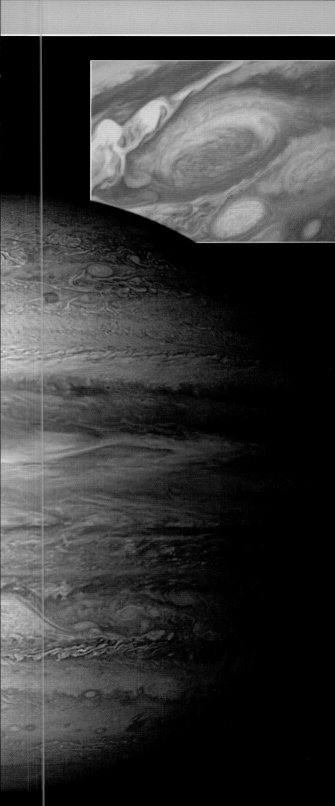

GREAT RED SPOT
Jupiter's atmosphere is dominated by storms whipped up by the planet's infrared heat and its rapid rotation. The Great Red Spot is a vast high-pressure storm that constantly changes in shape, size, and color, rotating counterclockwise every seven days.

Belts and zones
In a small telescope, Jupiter's bands of shading are readily discernible; a dark band of warm gas is called a belt, while a light band of cooler gas is known as a zone. The larger the telescope, the more apparent the differences become. There are two belts that straddle the equator that are always prominent: the north and south equatorial belts. The outlines of the belts, zones, and features within them are always changing, indicating vast, swirling disturbances in the atmosphere. Jupiter's fast rotation (completed in 10 hours) means you can watch a fixed point move across the planet in a single night. The most famous atmospheric feature is the Great Red Spot, which fills the width of the south-tropical zone and pushes into the south equatorial belt. This is a titanic, counterclockwise-rotating storm that has been raging for at least 350 years; it is so large that two Earths could very easily sit within it with room to spare. As the gases move and are heated from above and below, the Spot changes shape and color from almost invisible to flamboyant red.

OBSERVING JUPITER'S MOONS

Jupiter has more than 60 known moons, and the actual figure is probably nearer 100. Most are small chunks of rock and ice, so an observer using binoculars or a small telescope is unlikely to see more than the largest four: Io, Europa, Ganymede, and Callisto (in order of distance from Jupiter). These four are known as the Galilean satellites after the famous Italian astronomer Galileo Galilei, who first saw them in his new telescope in January 1610. They appear to us as faint points of light lining up on each side of the planet's equator, changing position as they orbit the planet. Sometimes one or more may be "missing," either passing in front of or behind Jupiter. The orbit of Io is only a little larger than that of our Moon around the Earth, and yet due to Jupiter's massive gravity, it travels once around the planet in just 1 day 18.5 hours (making this the length of its month). Io travels so fast, its movement can easily be seen over an hour.

JUPITER'S MOONS
This illustration shows the movement of the four Galilean satellites over one hour. In the top section, the moons running left to right are Io, Ganymede, Europa, and Callisto.

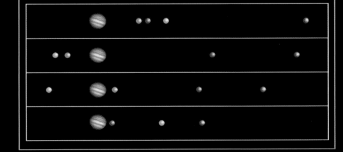

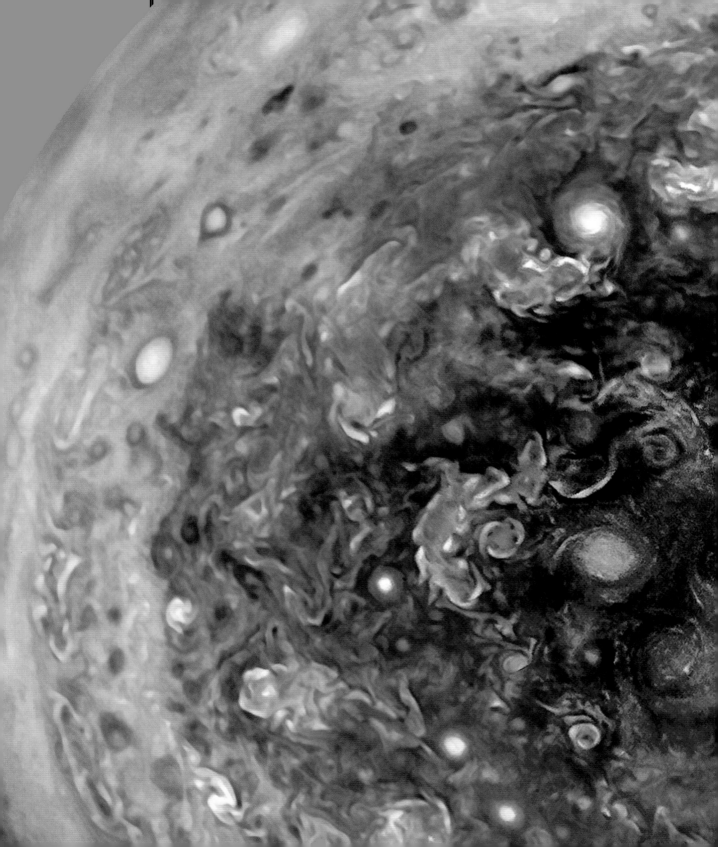

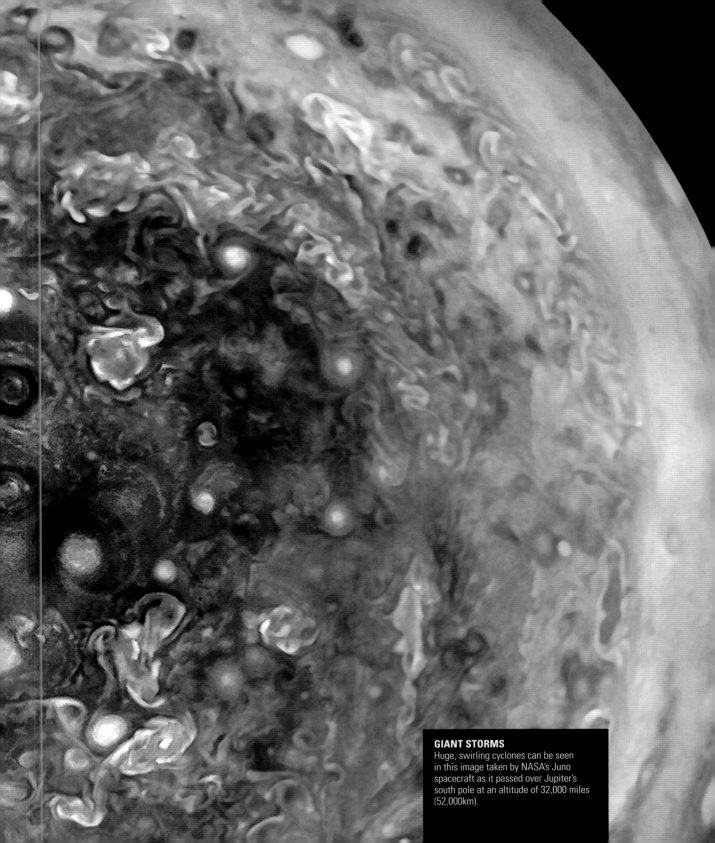

GIANT STORMS
Huge, swirling cyclones can be seen in this image taken by NASA's Juno spacecraft as it passed over Jupiter's south pole at an altitude of 32,000 miles (52,000km).

SATURN

Saturn is the second-largest planet in the Solar System. To the naked eye, it is a bright, creamy-colored disc; there is no hint of its rings until a telescope is trained upon this giant world.

Rings and moons

All the giant planets—Jupiter, Saturn, Uranus, and Neptune—have rings, but none are like Saturn's. Their rings are darker and less dense than Saturn's wondrous rings, made up of billions of large, bright, icy particles. These orbit the planet like tiny moonlets, reflecting the Sun's light to spectacular effect. From edge to edge, the rings stretch 174,000 miles (280,000km) through space, but their average thickness is only 0.6 mile (1km); they are very wide, flat, and thin. Their origin is unknown, but they may have existed since Saturn formed, changing over time as small moons and ring particles either smash apart in collision or merge together. Saturn also has at least 80 moons, one of which, Titan, is larger than Mercury and has its own atmosphere.

Distant orbit

As planets become more distant from the Sun, their orbital speed decreases, making them move more slowly across the sky. Saturn, orbiting beyond Jupiter, takes slightly over two years to travel though just one constellation. Opposition occurs every 54 weeks. At its brightest, to the naked eye Saturn appears as a slightly off-white star, but with a telescope of at least 25x magnification, it becomes one of the night's most rewarding sights. Even in a small telescope, the rings make the view truly memorable.

THE MOONS OF SATURN

With a small telescope, you may be able to spot a few of Saturn's main moons—in order of brightness, they are Titan (mag. 8.4), Rhea (mag. 9.7), Tethys (mag. 10.3), Dione (mag. 10.4), and Iapetus (10.1 to 11.9). The icy yet incredibly spongy and coral-like Hyperion, which measures 255 x 162 x 137 miles (410 x 260 x 220km), is the largest irregular-shaped object in the Solar System. It may be part of a larger object that broke up long ago. The much-smaller, oddly shaped Pandora is a "shepherd moon," which uses its gravity to keep Saturn's F ring in check.

SATURN'S MOON ENCELADUS

THROUGH A TELESCOPE
Saturn's surface shows little in the way of detail except some faint shading, but the ring system—2.25 times as wide as the ball of the planet—is what makes it stand out in the sky through even a small telescope.

SATURN PROFILE

MEAN DISTANCE FROM SUN	**886.7 million miles (1,427 million km)**
ORBITS THE SUN IN	**29.46 years**
ROTATES IN	**10 hours 40 minutes**
GREATEST MAGNITUDE	**-0.3**
FARTHEST FROM EARTH	**1,030.5 million miles (1,658.5 million km)**
NEAREST TO EARTH	**742.8 million miles (1,195.5 million km)**
GREATEST ANGULAR SIZE	**20.1 arcseconds**
NUMBER OF MOONS	**82**

SIZE COMPARISON

Saturn

Earth

The rings of Saturn

One of the most interesting things to observe is the changing aspect of the rings: over a period of about 7.5 years, our view changes from edge-on, when the rings are almost invisible, to a fully-opened 27° tilt, which allows us to see as much ring detail at opposition as possible. The greater tilt also reflects more sunlight our way, making the planet appear brighter. The planet itself shows very little in the way of features, mainly due to its small apparent size, but also because, unlike Jupiter, Saturn's atmospheric conditions do not produce marked color contrasts. The most detail an observer can expect is some slight dark and light banding in the atmosphere.

ROTATIONAL TILT
Saturn has a 27° axial tilt, so its north and south poles take turns to point toward the Sun. The tilt also affects our views of the planet and its rings.

Observing the rings

With a small-to-medium telescope, you may just be able to discern a definite dark split in the ring system, called the Cassini Division, which separates two of the main rings by 2,920 miles (4,700km). A larger telescope is essential to reveal the full seven rings, which are designated by letters: D, C, B, A, F, G, and E (working outward from the planet). The odd sequence comes from the order in which they were discovered. From Earth, we only see rings A, B, and C. Of these, B is the brightest and C the darkest, and after the Cassini Division, the C ring is the next easiest target. This faint ring (also known as the "crêpe ring" due to its lacy appearance) is best seen when it is in front of the bright planet disc. Within the A ring, look for the Encke Minima, a dark-looking band. Toward the outer part of the A ring, there is also a noticeable break, known as the Encke Gap. Other rings, gaps, and divisions are mostly too dark and faint to be visible from Earth.

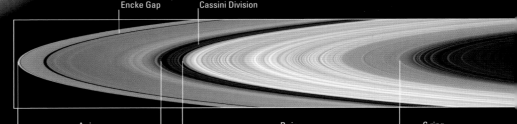

Encke Gap   Cassini Division

SATURN'S VISIBLE RINGS
This image was taken by the Cassini spacecraft, but a large amateur telescope will also show the A, B, and C rings.

A ring   B ring   C ring

URANUS AND NEPTUNE

In the dark, cold outer reaches of the Solar System lie the last of the planets, the cool, blue gas giants Uranus and Neptune.

Outermost planets

The two outer planets are very much a duo, similar in size, makeup, and appearance. Both are made chiefly of hydrogen and helium with water, ammonia, and methane, the latter being responsible for their blue color. Uranus is a featureless, murky green-blue ball, while Neptune is a translucent, crisp blue world with odd spots, storms, and wispy white clouds of frozen methane. Neptune has much greater wind speeds—at 1,240 miles (2,000km) per hour, they are the highest of any planet. Uranus is the bigger of the two, but its diameter only exceeds that of Neptune by 930 miles (1,500 km). Both have ring systems, but they are made of dark material, forming narrow and tenuous rings that are nothing like the great ring system of Saturn. Both also have a good selection of moons, visible through a powerful telescope.

FEATURELESS PLANET
In visible light, Uranus appears almost featureless with only vague cloud bands showing in even the best images.

URANUS PROFILE

MEAN DISTANCE FROM SUN	**1,784 million miles (2,871 million km)**
ORBITS THE SUN IN	**84.01 years**
ROTATES IN	**17 hours 14 minutes**
GREATEST MAGNITUDE	**5.5**
FARTHEST FROM EARTH	**1,962 million miles (3,157 million km)**
NEAREST TO EARTH	**1,604 million miles (2,582 million km)**
GREATEST ANGULAR SIZE	**3.7 arcseconds**
NUMBER OF MOONS	**27+**

SIZE COMPARISON

Earth

Uranus

THROUGH A TELESCOPE
Even through a telescope Uranus will only show as a rather disappointing, subtle greenish disc. The planet is so remote that the amount of light it receives from the Sun is only 0.25 percent of that on Earth.

Discovering remote worlds

Both Uranus and Neptune owe their discovery to the invention of the telescope. Uranus actually became the first planet to be discovered by telescope, when the astronomer William Herschel found it using his telescope observing from the garden of his home in the town of Bath, UK. Many years of observing Uranus revealed irregularities in its orbit. The explanation could be the gravitational pull of a further world. The hunt was on for planet number eight, and in 1846 Neptune, the first planet discovered by mathematical means, was found in the constellation Aquarius.

URANUS'S AXIAL TILT

The most unusual aspect of Uranus is that its axis is tilted so much that it appears to move in a similar way to a ball rolling along the ground. By comparison, the Earth's axial tilt is 23.5 degrees, Jupiter's is just 3.1 degrees, while Neptune has a 29.6-degree "lean." The tilt of Uranus at 97.86 degrees is quite dramatic and probably arises from the planet having had a major collision with another object in some ancient epoch. Only one planet has a greater axial tilt and that is Venus. According to one theory our neighboring world must have been knocked so hard as to virtually completely turn it over, for Venus's tilt is 177.3 degrees.

SPINNING ON ITS SIDE
The rotation axes of all the planets are tilted to some degree, but so much so in Uranus that its equator and surrounding faint ring system appear almost vertical.

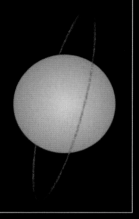

NEPTUNE PROFILE

MEAN DISTANCE FROM SUN	2,794 million miles (4,497 million km)
ORBITS THE SUN IN	164.79 years
ROTATES IN	16 hours 7 minutes
GREATEST MAGNITUDE	7.8
FARTHEST FROM EARTH	2,907 million miles (4,678 million km)
NEAREST TO EARTH	2,676 million miles (4,306 million km)
GREATEST ANGULAR SIZE	2.3 arcseconds
NUMBER OF MOONS	13+

SIZE COMPARISON

Earth

Neptune

Observing Uranus and Neptune

With increasing distance from the Sun, the amount of sunlight available to reflect decreases, making observation of these two far-off worlds challenging. With its 84-year orbit, Uranus travels incredibly slowly across the sky, taking 7 years to traverse a single constellation. Neptune is more than 1,600 million miles (2,600 million km) further from the Sun than Uranus, and has an orbit of 165 years—it takes 15 years to cross one constellation. Both planets are at opposition roughly once a year, but because these planets are so distant anyway, this makes only a few tenths of a magnitude difference to their brightness. At its brightest Uranus has a magnitude of 5.5, just visible to the naked eye as a faint "star." Binoculars will find the planet easily, but a good-sized telescope is needed to see it as anything more than a point of light. Neptune's maximum magnitude is only 7.8, beyond the limits of the naked eye, but even modest binoculars will be able to locate it, and a telescope will reveal the disc of the planet.

WINDY WORLD
Close-up images taken by Voyager 2 revealed Neptune's fast-moving weather systems.

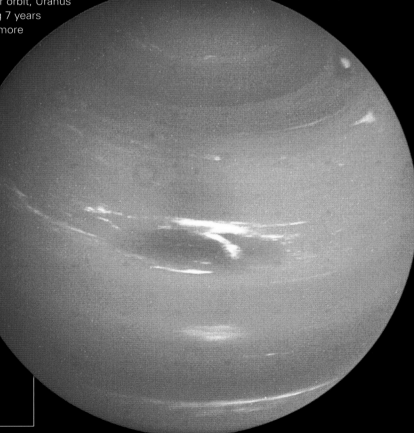

THROUGH A TELESCOPE
Neptune is smaller than Uranus, but the vivid blue of its atmosphere makes it stand out more strikingly against the blackness of space.

DWARF PLANETS AND ASTEROIDS

In addition to the eight planets, the Sun is circled by countless smaller objects, the largest of which are classed as dwarf planets and asteroids.

The dwarf planets

From its discovery in 1930 until 2006, Pluto was the ninth planet of the Solar System. However, by the end of the second millennium it had become clear that Pluto is not alone out in the far reaches beyond Neptune; it is one of a remote belt of minor bodies, and some of these are as large as if not larger than Pluto. In order to limit the number of objects designated as planets, the International Astronomical Union defined their characteristics, along with those of dwarf planets and asteroids. The result was that the Solar System now officially contains eight planets, and five dwarf planets. The two most significant dwarf planets—Pluto and Eris—are very far away, while Ceres (formerly classed as the largest of the asteroids) is in a belt of rocky objects that orbit the Sun between Mars and Jupiter.

PLUTO
A color-enhanced image of Pluto shows the dwarf planet's surface in great detail, including large areas of nitrogen, carbon monoxide, and methane ices.

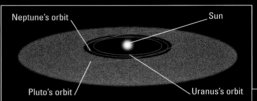

EDGEWORTH–KUIPER BELT
Sometimes referred to as the Solar System's final frontier, the Edgeworth–Kuiper Belt lies 2.8–4.6 billion miles (4.5–7.5 billion km) from the Sun. It contains more than 35,000 objects greater than 60 miles (100km) wide, including the eccentrically orbiting Pluto..

Belts of minor worlds

Beyond Neptune, a swarm of icy bodies known collectively as the Edgeworth–Kuiper Belt orbits the Sun. It is similar to the inner Asteroid Belt, but it is more distant, much larger, and contains many more objects. The small bodies in the Asteroid Belt are mainly composed of rock and metal, but those in the Edgeworth–Kuiper Belt are icy worlds kept from melting by their vast distance from the Sun.

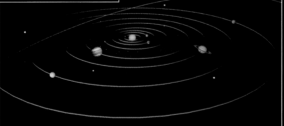

ASTEROID BELT
The Asteroid Belt lies between Mars and Jupiter. It contains billions of space rocks of irregular shapes.

Ceres, queen of the Asteroid Belt

Lying between the orbits of Mars and Jupiter is the Asteroid Belt, home to hundreds of thousands of small worlds composed mainly of rock, metal, or a mixture of both. Their presence may be due to the massive gravity of nearby Jupiter. There are so many objects here that they are liable to collide with one another, causing their irregular shapes. Some of the chips of rock formed during these encounters have traveled down to Earth in the form of meteorites, which is how we know their chemical constituents. Since the redesignation of Ceres as a dwarf planet, Vesta is now the largest of the asteroids. Even though its diameter is about half that of Ceres, Vesta appears brighter in the sky due to its much lighter surface.

CERES
Its diameter of 584 miles (940km) and its nearly spherical body earned Ceres its "promotion" from asteroid to dwarf planet.

Pluto and the Plutoids

Astronomers are aware of more than 1,000 objects belonging to the Edgeworth–Kuiper Belt, of which Pluto is the brightest. These are often described as Trans-Neptunian Objects (TNOs), though Pluto itself sometimes crosses inside Neptune's orbit. The TNOs designated as dwarf planets are called Plutoids. One of these—Eris, discovered in 2006—is actually larger than Pluto itself. It only has one confirmed moon, named Dysnomia (which is the third-largest object in the Edgeworth–Kuiper Belt), while Pluto has three moons. As of January 2010 there were four confirmed Plutoids: Pluto, Eris, Haumea, and Makemake, with a further ten or so possible candidates, including Quaoar, Orcus, Ixion, and Sedna. Astronomers are now investigating how and where these dwarf planets were formed.

ASTEROID 243 IDA

Ida is a typically irregular asteroid that is around 36 miles (58km) long and 14 miles (23km) wide. Its numerous, degraded craters suggest that it is not a youthful object.

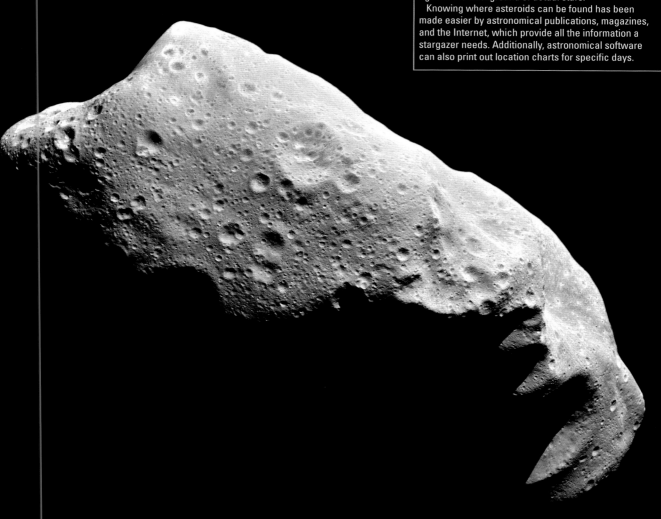

THE SUN

The centerpiece of the Solar System is a truly fascinating object, but rigorous safety precautions are essential for observing it.

The brightest star

Astronomy is an interest so associated with the night that it is easy to forget that, during the day, we are able to see a blazing star that is virtually on our doorstep. However, viewing this star is by no means as easy as gazing at the constellations, because we cannot look at it directly. This massive ball of incandescent gas has an incredible magnitude of -26.7, and is far too bright to be viewed by the naked eye without damage to the retina. **Warning:** never look at the Sun with the naked eye, or through binoculars or a telescope; the concentration of light gathered in through the eyepiece can cause immediate blindness.

Observing the Sun safely

Solar telescopes, incorporating filters that enable direct viewing of the Sun, are available (see p.54), but these are specialized pieces of equipment. There are also add-on solar filters for telescopes and "solar binoculars" on the market; these need to be used with caution to avoid incidental light. With an ordinary, small refracting telescope, the safest way to view the Sun is by a method called projection. This entails directing the Sun's image through the eyepiece onto a piece of white cardboard. The further the card is from the eyepiece the larger, but fainter, the Sun's image will be.

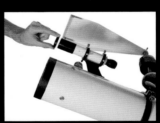

1 CAP THE LENS FOR SAFETY
Never look for the Sun in the telescope by eye. For safety, place a cap over the finderscope. If you are using a refractor, also cap the main lens.

2 FINDING THE SUN
To aim the telescope at the Sun, move it around until you locate a position in which the telescope tube or finder casts the shortest shadow.

3 SHARPENING THE IMAGE
Remove the lens cap (if using a refractor) and focus the Sun's image onto a piece of white cardboard. Adjust the alignment and focus for sharpness.

OUR LOCAL STAR

The Sun is our closest star and it is a fine example of a giant hydrogen and helium sphere that is perfectly balancing its counteracting forces. The hot gas would expand out into space if it were not counteracted by the effect of gravity—the weight of layer upon layer of gas pressing inward. It has to be perfectly balanced, because stars that are unstable can change wildly in brightness, size, and energy output in a way that would make life on Earth difficult, if not impossible.

The Sun, together with all the other objects that make up our Solar System, was formed out of a cloud of dust and gas (a nebula) around 4.7 billion years ago. It is an ordinary main sequence star (see p.16–17) classed as a yellow dwarf. Now about halfway through its life, it will eventually expand into a red giant, but there are several billion years to go before the Sun's aging produces any noticeable effects on our planet.

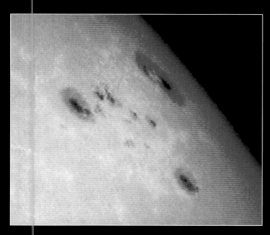

The light from a star

When we look at the Sun, with adequate safety precautions, what we are actually seeing is its photosphere (light sphere), which has a temperature of just under 10,000°F (5,500°C). Embedded in the photosphere are sunspots, which are lower-temperature areas of the photosphere. These have lighter and darker areas (of penumbra and umbra) and are often found in groups. They appear and disappear cyclically, lasting anywhere from a few days to a few weeks. They can be seen to move day by day as the Sun rotates over about a month. The Sun's surface granulation can be seen using a large-aperture specialized telescope. These mottled patches are pockets of hot gas which have risen to the Sun's surface, and they last around 20 minutes. The outer part of the Sun's atmosphere, the corona, can be seen when the disk of the Sun is blocked off during a total solar eclipse.

THE SURFACE OF THE SUN
Sunspots appear as darker patches on the Sun's surface. They are signs of solar activity, where the Sun's magnetism is disturbed.

Solar eclipses

A solar eclipse occurs when the Moon blocks sunlight from reaching Earth. It is caused by the alignment of three Solar System bodies: the Sun, Moon, and Earth. The Sun, of course, is the largest of the trio, being 400 times larger than the Moon. By a strange chance it is also 400 times further away, which means the Sun and Moon look almost exactly the same size in the sky. This allows the Moon to just about cover the Sun, in a few minutes of "totality," revealing the glorious sight of the corona. If you are standing just outside the path of totality (beyond the path of the Moon's central shadow, or umbra), you will witness only a partial eclipse, where only a part of the Sun is covered. Total eclipses are extremely rare at any one place on Earth, but partial eclipses are more widely visible.

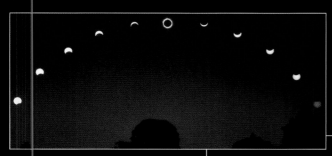

ECLIPSE OF THE SUN
A multiple-exposure photograph (above) captures the stages of a total eclipse, including the moment of total eclipse (the central point), when only the corona of the Sun is visible. At the beginning and end of the eclipse, the rough, cratered surface of the Moon allows small amounts of sunlight to break through, known as "Baily's Beads" (right).

THE PATH OF TOTALITY
The Moon is much smaller than the Sun, so its shadow during an eclipse is cone-shaped. It consists of the central umbra (associated with the path of totality across the Earth) and the penumbra (area of partial eclipse).

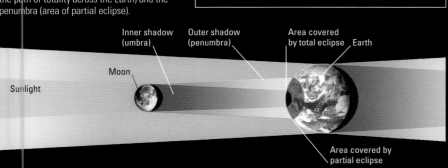

Inner shadow (umbra)
Outer shadow (penumbra)
Area covered by total eclipse   Earth
Moon
Sunlight
Area covered by partial eclipse

ECLIPSE VIEWER
A safe and easy-to-make eclipse viewer consists of two pieces of cardboard, one of which has a small hole at the center. Allow the sunlight through the hole of the top card, and its image will appear on the second one held underneath. Vary the distance between the cards to obtain the sharpest view.

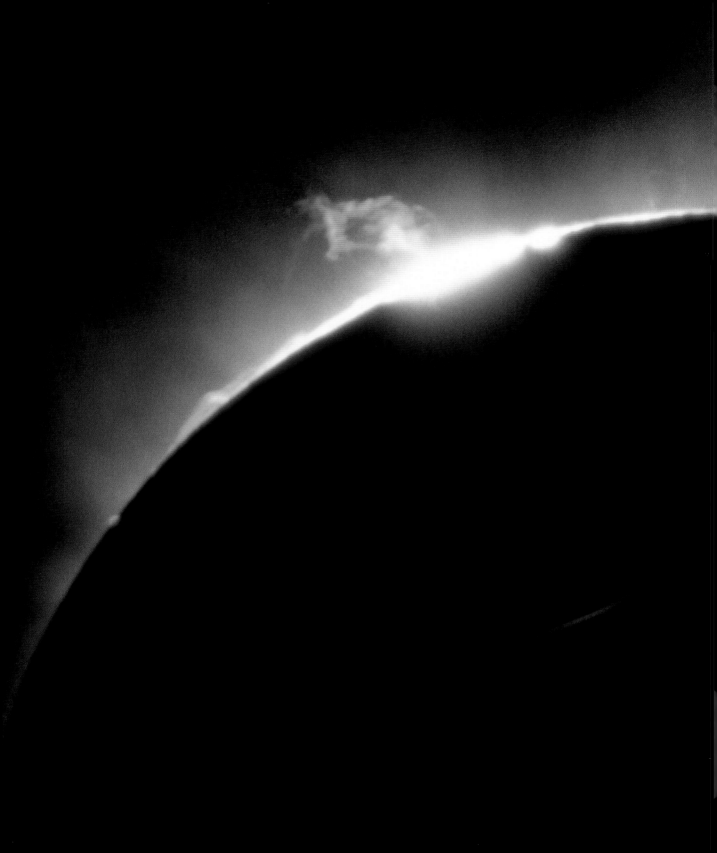

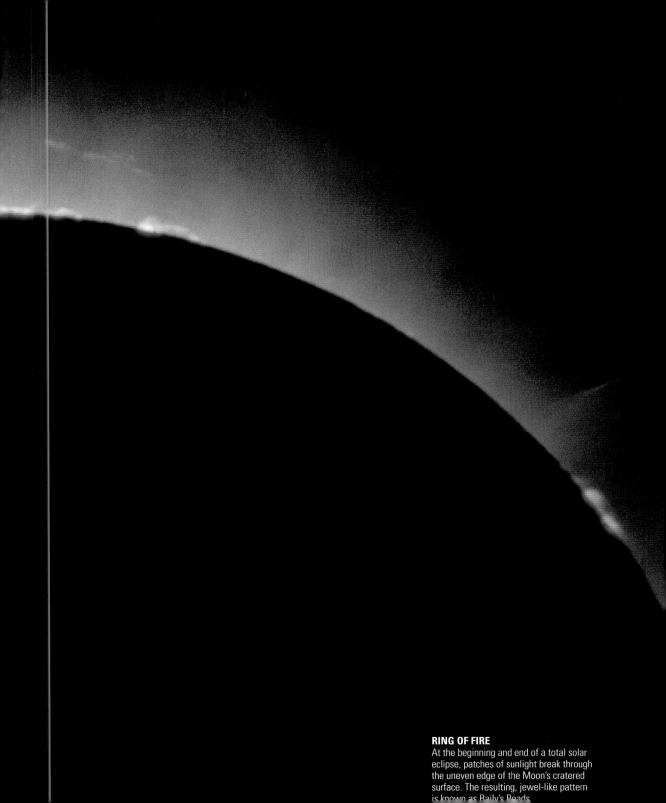

RING OF FIRE
At the beginning and end of a total solar
eclipse, patches of sunlight break through
the uneven edge of the Moon's cratered
surface. The resulting, jewel-like pattern
is known as Baily's Beads.

THE MOON

**The Moon is the Earth's only natural satellite.
It is so close to Earth that its surface features
are discernible even to the naked eye.**

Our closest neighbor

The Moon orbits fairly close to Earth, at a distance of only 238,900 miles
(384,400km). At about a quarter of Earth's size, it is large enough for Earth
and Moon to be considered a double planet. Like all moons and planets, it
emits no light of its own—moonshine is simply reflected sunlight. The Moon
is a wonderful object to observe; binoculars alone will reveal its broad
plains, countless craters, deep valleys, and jagged mountain ranges.

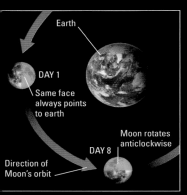

Earth

DAY 1

Same face
always points
to earth

Moon rotates
anticlockwise

DAY 8

Direction of
Moon's orbit

SYNCHRONOUS ROTATION

The Moon rotates like the Earth, but at a much
lower rate. Its orbit around us takes 27.3 days,
and that is the exact length for its rotation too.
The result is that the Moon always shows us
the same side; we can never see the far side.

The Moon's orbit

We see the different phases of the
Moon as it travels around us. At new
Moon it lies between Earth and the
Sun. If the alignment is perfect an
eclipse of the Sun will result (see
p.195), but the Moon's tilted orbit
means that for most of the time it
travels unnoticed above or below the
Sun. After a couple of days, the
Moon is seen as a thin crescent in
the western evening skies, following
the Sun down shortly after sunset.
Day by day the Moon moves further
east, away from the Sun, its phases
waxing through half-Moon, then
gibbous, and finally full Moon, which
occurs 14½ days after new Moon. At
full Moon it is halfway round its orbit,

and here too, if the alignment is perfect, there will be an eclipse—but this
time, of the Moon. Although the Moon travels around us in 27.3 days, the
phase cycle takes 29.5 days because the Earth has itself also moved in space
during the cycle, changing the angle from which we see the sunlit Moon.

WAXING CRESCENT   FIRST QUARTER   WAXING GIBBOUS   FULL MOON   WANING GIBBOUS   LAST QUARTER   WANING CRESCENT   NEW MOON

MOON PHASE

The Sun, of course, is always lighting up one half of the Moon; the amount of
lighted side we can see depends on whereabouts the Moon is in its orbit around
Earth. The Moon is said to be waxing when its phases are growing from an evening
crescent until full Moon, and waning afterward as it journeys back to new Moon.

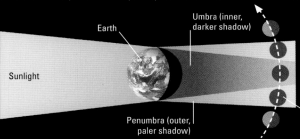

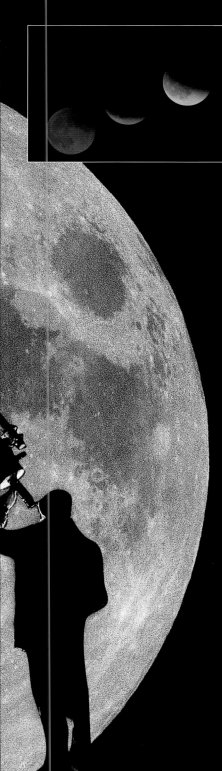

Lunar eclipses

Lunar eclipses are not as spectacular as solar eclipses, but they are still an amazing sight. They occur when the Earth lies between the Moon and the Sun, and the plane of the Moon's orbit coincides with the plane of the Earth's orbit around the Sun—which happens several times a year, always during a full Moon. The first sign of a total lunar eclipse is a darkening on the eastern side of the Moon. Over the next few hours the darkness slowly sweeps across the lunar disc as the Moon continues to slip into the Earth's shadow (the outer penumbra). As the Moon approaches the exact opposite side of the sky to the Sun, at the moment of full Moon, it enters the deepest, darkest section of Earth's shadow (the umbra), and takes on a reddish hue, before eventually the shadow starts to slip from the disc as the Moon moves out through the fainter penumbra. Sometimes it is quite a surprise how much the sky darkens during totality, when the Moon is fully eclipsed. The fainter stars, normally washed out by reflected sunlight, become visible in the darker skies that surround the now reddish-brown Moon. The whole scene can stay this way for over 90 minutes, which, unlike a total solar eclipse, allows plenty of time to observe the event.

THE RED MOON
At the point of total eclipse there is no direct light falling on the Moon, but it appears red, not black, due to the indirect sunlight still being scattered toward it from Earth's atmosphere.

Sunlight

Earth

Umbra (inner, darker shadow)

Penumbra (outer, paler shadow)

PATH OF A LUNAR ECLIPSE
In a total eclipse, the full Moon moves through all parts of the Earth's shadow, from the weaker penumbra (outer shadow) to the intense inner umbra and then the penumbra again.

Only a slight darkening of the Moon occurs in the light outer shadow

FORMATION OF EARTH'S MOON

One of the most widely accepted theories of how the Moon came to be created suggests that a large object (about the size of Mars) collided with the young Earth. This produced a vast amount of molten material, partly from the upper layers of the Earth but also from the impactor, which was then splashed into orbit around the Earth. Gravity slowly clumped this material together and the Moon came into being. Since then the Moon has been moving away from us, at a rate of around 1½ inches (4cm) per year. This is down to gravitational tidal effects, as the Moon collects energy from the Earth-Moon system allowing it to move away, while causing Earth's rotation to slow down. But this is not dramatic, as each day is only lengthening by about 2.2 seconds every 100,000 years.

1. COLLISION
It is thought that the Moon was formed when a large asteroid hit the young Earth. A huge amount of molten, silicate material was ejected from the Earth's mantle, which formed a massive cloud of gas, dust, and rock. The heat radiated away, and the cloud began to cool.

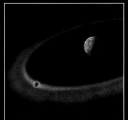

2. DEBRIS COALESCES
The majority of the ejected material from the impact settled into a circular orbit around the Earth, forming a dense, doughnut-shaped ring. The rocks within the ring began to collide with one another, until a single body dominated the ring. This became Earth's only satellite—the Moon.

Observing the lunar surface

The Moon's surface bears the scars of constant bombardment and upheaval that took place during its formation. The naked-eye view of the Moon shows at first a gray world of light and dark patches, but as you become accustomed to viewing, you should be able to see the sites of some of the biggest impacts. These are "ray craters," made when objects smashed into the Moon, sending out great splashes of molten rock. The rays most easily discerned by the naked eye surround the craters of Copernicus and Tycho. Binoculars reveal the craters themselves, and define the dark and light landscape features more clearly, allowing you to identify the various "seas" or maria on the surface, but to immerse yourself in this compelling world, telescope viewing is a must.

The maria are vast **lava flows** of basaltic material like that which erupts from **Earth's volcanoes**.

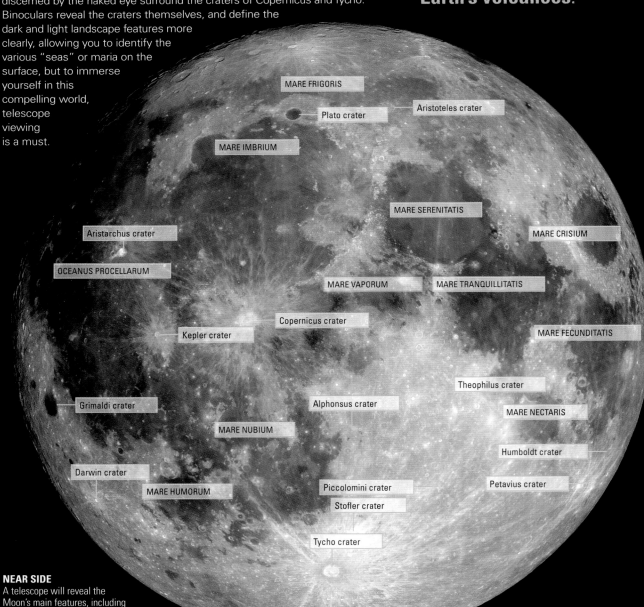

MARE FRIGORIS

Plato crater

Aristoteles crater

MARE IMBRIUM

MARE SERENITATIS

Aristarchus crater

MARE CRISIUM

OCEANUS PROCELLARUM

MARE VAPORUM

MARE TRANQUILLITATIS

Copernicus crater

Kepler crater

MARE FECUNDITATIS

Theophilus crater

Grimaldi crater

Alphonsus crater

MARE NECTARIS

MARE NUBIUM

Humboldt crater

Darwin crater

Petavius crater

MARE HUMORUM

Piccolomini crater

Stofler crater

Tycho crater

NEAR SIDE
A telescope will reveal the Moon's main features, including its many craters and its dark lowland plains (the "seas" or maria), which contrast with the brighter highlands.

When to observe

Although the Moon always shows the same face to Earth, its features change in appearance through the month as sunlight strikes them from different angles. Viewing at full Moon can be disappointing, as its surface is most evenly lit by the Sun and its features are indistinct; the weeks either side of the new Moon are best. As the Moon waxes or wanes you can look along the "terminator"—the line that divides the Moon's sunlit and dark sides. Here there are always long shadows, which throw landscape features into sharp relief.

THE TERMINATOR
Sunlight strikes the Moon at its narrowest angle at the terminator, where light meets dark. This creates a high-contrast view, and reveals the greatest amount of detail.

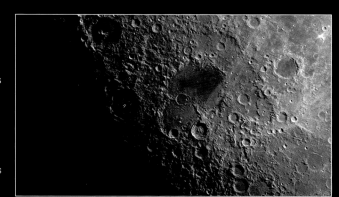

LANDSCAPE FEATURES OF THE MOON

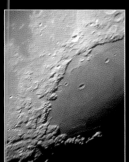

SEAS (MARIA)

The ancients believed the dark areas of the Moon were seas—in Latin, *maria*—set against the lighter land. Today we know there is water on the Moon trapped in the dark, unlit bases of craters around its poles, but the maria are just smoother rocky areas or plains caused by the upwelling of molten rock long ago that flooded large portions of the lunar surface. This makes the maria younger than the bright highlands, something confirmed by the fact that the maria have far fewer craters than the highlands.

ESCARPMENTS (RUPES)

These straight features are caused by a long vertical shift of the Moon's crust along a fault. The land on one side is either pushed up or slips down, and the result is an angled cliff. Probably the best example is the Straight Wall, or Rupes Recta, which runs for 78 miles (125km) through Mare Nubium. The longest escarpment of all is Rupes Altai at 265 miles (427km) long. This is the surviving section of a ring that is thought to have surrounded the Necaris Basin (a huge impact crater) soon after its formation.

MOUNTAINS (MONS)

These can be solitary peaks or chains, some with recognizable names, such as Mont Blanc in the (lunar) Alps. Away from the maria the Moon is covered in mountain highlands, but there are mountains in the maria too—just like islands on Earth that are really just the tops of vast submerged mountains, there are one or two peaks visible that were too high to be covered by lava when the maria formed. The highest mountain is the 3-mile (4.7-km) high Mons Huygens, found in Montes Apenninus.

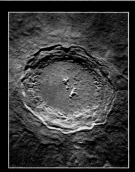

IMPACT CRATERS

Impact craters are formed when objects smash into the Moon. What happens next depends on many factors including the size of the object, but generally the energy released causes the surface of the Moon to melt, together with some or all of the impactor. The molten material then "splashes" outward until it cools to form the outer rim of the crater. If there is enough energy and the rock is still molten enough, it can bounce back to meet in the middle, where the molten rock solidifies to form a crater peak.

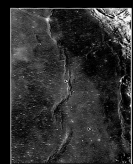

WRINKLE-RIDGES (DORSA)

These gently-winding low ridges in the maria formed as the molten rock that made the maria cooled and contracted. Some give a hint of hidden features below the surface. These ridges can stretch for hundreds of miles across the Moon's surface, but they are no more than a few hundred yards high. They are easiest to see very close to the terminator, as the low sunlight lets them cast more shadow. The longest ridge is Dorsa Geikie, which stretches for 142 miles (228km) in Mare Fecunditatis.

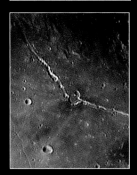

RILLES (RIMA)

Rilles are the exact opposite of wrinkle-ridges (above). They do not rise above the surface, but cut down into it much like a trench or the groove in a vinyl record. Rilles are much wider, however, than ridges, with widths measured in kilometers. They are formed where the land has sunk, perhaps because the roof of an old, long lava tube has collapsed, and as a result their lengths can reach well over 60 miles (100km). The longest rille is Rima Hesiodus at 159 miles (256 km) on the rim of the crater Hesiodus.

COMETS AND METEORS

When particles released from frozen, dusty comets hit Earth's atmosphere, they produce meteors—our "shooting stars."

Dirty snowballs

Comets are great chunks of ice and dust that orbit the Sun, often on long, looping paths that take them far into interstellar space. As a comet nears the Sun, it starts to vaporize in the heat, producing a dramatic spectacle. During their journey around the Sun comets also continually shed tiny particles of rocky dust—meteoroids—which become meteors (shooting stars) when they enter Earth's atmosphere. Thousands of tons of dust from interplanetary space enter Earth's atmosphere each year.

STREAKING COMETS
Halley's comet displays the distinctive round coma and streaming tail. We can see comets because they reflect the Sun's light, and their gases release energy absorbed from the Sun.

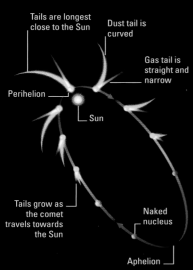

Tails are longest close to the Sun

Dust tail is curved

Gas tail is straight and narrow

Perihelion

Sun

Tails grow as the comet travels towards the Sun

Naked nucleus

Aphelion

Where do comets come from?

Comet nuclei range from 0.6 to 30 miles (1 to 50km) across and are thought to be remnants from the formation of the Solar System. When the Sun finally began to shine, its radiation pushed lighter material beyond the orbits of the planets, into the Edgeworth–Kuiper Belt (see p.190). This is thought to be the home of the "short-period comets"—those with orbits up to about 200 years. Much further out there is thought to be a vast repository of comets called the Oort Cloud, in the form of a halo surrounding the Solar System. This is the source of long-period comets, which take thousands of years to complete an orbit. If their paths are diverted to cross the orbits of the planets, they may become trapped in much smaller orbits or destroyed in a collision. Any of the short- or long-period comets may be nudged by the gravity of a planet or passing star and sent hurtling towards the Sun, to make the nucleus grow into a bright coma and tail, lighting up the night skies.

STARS WITH TAILS
When a comet approaches the Sun it vaporizes, releasing trapped dust that disperses to form a bright head, or "coma." Radiation from the Sun pushes the dust particles away from the coma, while the solar wind converts some of its gases into ions, creating the comet's two tails.

OBSERVING COMETS

The great long-period comets, such as Hale–Bopp (left), will not return for thousands of years, but there are several short-period comets to observe each year, the dates for which are published in astronomy magazines. Most are very faint and you usually need binoculars at least to see them, and even then you may only see a tail. However, new comets are always being discovered, and astronomical internet sites react quickly to breaking comet news. In most cases, the comet will look like a fuzzy patch of light; you will not be able to see it move, but you can chart its nightly progress against the starry background. The technique of averted vision (see p.48) is particularly useful for viewing comets.

OBSERVING A METEOR SHOWER

The time and location of the known annual meteor showers is entirely predictable; a table listing the main ones appears on p.237. They are best viewed with the naked eye. Find as dark a site as possible with a wide sky, and look toward the radiant constellation. Necks can start to ache after looking up for too long, so a reclining garden chair is recommended. Light pollution or moonlight will reduce the numbers of meteors you see. If the Moon is gibbous or full on the peak dates for the shower, its light may wash out many of the fainter meteors; a new Moon provides perfect conditions.

Larger meteroids—the size of a **marble**—appear brighter than **all the stars and planets** around them.

Meteors and meteor showers

Shooting stars, or meteors, can be seen on any clear, dark night. Each tiny particle that generates such an event is simply debris moving through space when Earth gets in the way. The particle hits our atmosphere at a speed of anything up to 46 miles (74km) per second, where it vaporizes and the resulting trail is seen as a silent, glowing streak. Odd occasional meteors, called "sporadics," are caused by wandering meteoroids (probably left behind by some long-gone comet) and can appear anywhere in the sky at any time. However, there are dates in the year when meteor showers occur. These appear from the same point, known as the radiant, each year, and the constellation that holds the radiant gives the name to the shower. So the radiant in Lyra is where the Lyrid meteor shower, or simply the Lyrids, appears. Most meteor showers result from Earth crossing a stream of debris that has accumulated in the orbit of a comet, where it may persist for centuries. As the Earth enters this region each year, dusty particles hit the atmosphere. A good annual event such as the Perseids (caused by dust from comet Swift–Tuttle, which last passed Earth in 1992) can produce 90 or more meteors per hour.

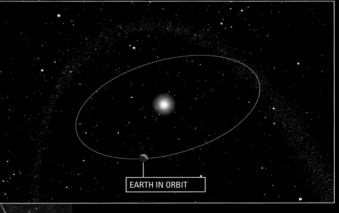

EARTH IN ORBIT

THE ORIGINS OF A METEOR SHOWER

When Earth passes near a comet's orbit (the dust trail in the image above), it sweeps through debris left behind by the comet, and each tiny particle that hits Earth's atmosphere turns into a meteor.

ATMOSPHERIC PHENOMENA

Not all of the interesting targets for observation lie in space. There are many fascinating phenomena in the sky that the Earth "creates" itself.

Earthly light effects

When light or charged particles from the Sun reach Earth's atmosphere, either directly or, in the case of light, by way of reflection from the Moon, conditions sometimes conspire to produce beautiful effects for observers on the ground. Some of these phenomena, such as aurorae (the Northern or Southern Lights) can only be seen from certain places or at certain times; some need perfect weather conditions, but others can appear anytime, anywhere.

MOON HALO
A Moon halo is caused by hexagonal ice crystals in the atmosphere diffracting moonlight into a circle around the Moon at a radius of 22°. The ice crystals are not all aligned, so from the observer's point of view they form a ring all around the object producing the light.

SUN DOGS
Sun dogs appear as bright patches 22° or more on either side of the Sun. Though caused by ice crystals high in the atmosphere, they are often seen even when the weather is not cold. Here they are part of a more extensive display.

Sun and Moon dogs

Sun dogs, or parhelia, are bright patches that usually appear in pairs, one each side of the Sun. They are caused by the diffraction of the Sun's light by neatly arranged hexagonal-shaped ice crystals in cirrus clouds, and they occur fairly often. If the ice crystals have not been perfectly aligned, sun dogs may sit on a halo around the Sun (like the Moon halo shown above). Rarely, another halo may be visible that circles the Sun horizontally, also passing through the sun dogs. This is known as a parhelic circle. Paraselenae, or Moon dogs, are much less common than parhelia, as the Moon has to be full or near-full in order to reflect enough light into the atmosphere, but once seen they are never forgotten.

Noctilucent clouds

Noctilucent (night-shining) clouds are thin, wispy white clouds made of ice crystals that appear after sunset at altitudes of around 46–52 miles (75–85km). Invisible during the day, when their faint light is washed out by the bright sky, they begin to shine once the Sun has disappeared but while its light is still illuminating the upper atmosphere. They are visible in both hemispheres at latitudes between 50° and 65° north and south during the summer months. The particles around which the ice crystals develop are thought to have come down from space as micrometeoroids, and up from volcanoes and dust.

NOCTILUCENT CLOUDS
Noctilucent clouds are seasonal, and may appear from May to August in the northern hemisphere, and November to February south of the equator.

Aurorae

This magical spectacle is caused by electrically charged particles being funnelled down toward the polar regions when a blast of particles from the Sun disturbs Earth's magnetic field. There they excite gas atoms in the atmosphere into a shimmering display known as Aurora Borealis, or Northern Lights, in the northern hemisphere, and Aurora Australis, or Southern Lights, in the southern hemisphere. The light show itself can tower as high as 200 miles (300km) or more above Earth's surface. The colors that are seen depend on the incoming particles' energy and which atmospheric gas they excite. An auroral display can last for many hours. Quite often it starts with a low, soft green arc, growing in brightness and size before red and green rays appear, which then ripple out into long ribbons that stretch across the sky. In a trick of perspective, the rays of an auroral corona seem to converge right above your head.

OTHER LIGHTS IN THE SKY

Much goes on in the sky, day and night, and sometimes the great distances involved make it difficult to tell exactly where or what an object is.

Unidentified sky objects

If you see a bright light in the sky that you don't recognize, the first question to ask is whether the object is moving. This can be ascertained by watching it for a few minutes (although if it is moving, it is likely to be obvious within seconds). If it is stationary, the object is likely to be a planet. At its brightest, Venus is far brighter than any star, and it is often mistaken for a UFO. If you see the object within three hours of sunrise or sunset, it's even more likely to be Venus, but consider also Mars or Jupiter, which are less bright than Venus but can nonetheless look like bright stars. If you observe a similarly bright object that is moving, check whether there is an airport nearby. Red and green flashing lights or associated noise would indicate a plane on approach to landing.

VENUS
Venus is the brightest of all the planets, and at sunset it appears low on the western horizon. Its strong light beaming through trees in a sky still washed by sunlight often confuses observers.

The International Space Station

There is one other object that can move and appears bright, but this will always be seen traveling at a constant, steady pace across the night sky. It emits no audible sound, as it flies in space about 366 miles (590km) overhead. This is the International Space Station (the ISS). This craft has been getting brighter since the first section was placed in orbit in 1989, as each new additional module increased the amount of sunlight the craft reflected. But it was not until the final set of solar panels were added in 2009 that the station became bright enough to be visible in daylight.

PATH OF THE ISS
This long-exposure photograph shows the ISS traveling through the night sky. It can be seen in daylight, if its exact position is known at the time.

IRIDIUM SATELLITES

Iridium satellites form a network used by communication devices. There are 66 such satellites circling the globe at a higher altitude than the ISS. Iridium satellites are visible day and night, but these are intensely bright and quite fast. Suspect one if you see a moving point of light that gradually brightens to a dazzling magnitude before fading away within ten seconds or so—this sudden reflection of light is known as an iridium flare. Details of where and when Iridium flares will occur can be found at: www.heavens-above.com.

THE BRIGHTEST SATELLITES
Iridium flares are due to the array of antennas on the satellites acting like a perfect door-sized mirror, directly reflecting light from the Sun down to an exact location on Earth.

The planet **Venus**, meteors, fast-moving satellites, and even **ice crystals** have been mistaken for UFOs.

Weather balloons and lanterns
There are other lights in the sky that may behave more erratically, such as moving in one direction and then the next. If it is daytime or just after sunset, these lights could turn out to be weather balloons, made of shiny silver material called Mylar. These objects reflect the Sun's light, so their rise and fall or any other odd movement can be quite perplexing for the unknowing observer. At night, strange bobbing lights may be luminous Chinese lanterns. Both weather balloons and lanterns are at the mercy of the atmosphere, and as lightweight objects traveling upward they will probably encounter winds moving in different directions. For example, a Chinese lantern may be observed as a light climbing gently, before suddenly taking a different course (as a high atmospheric wind shears its path), before suddenly disappearing completely (as the wind blows the lantern out). All this can appear as though a UFO has lifted off and is departing for some distant planet, and may account for many "alien craft" sightings.

Satellites
Satellites can be seen each night, resembling faint stars that cross the sky in minutes. Their apparently slow movement is due to their great distance from us; in reality most are speeding around the Earth at about 16,800 miles per hour (27,000kph), taking only about 90 minutes to complete one orbit. Just as with the International Space Station, it is only possible to see a satellite because it reflects light from the Sun; none has lights of any sort. If a satellite enters or exits the Earth's shadow along its orbit—which often happens as you observe one—the Sun's light is not instantly cut off as you watch the satellite disappear, but fades over 5 to 10 seconds.

SPACE HARDWARE
The ISS is a working laboratory orbiting Earth at 17,500mph (32,400kph). Its solar panels exceed the wingspan of a Boeing 777 airliner and provide enough power for all the station's components and experiments.

WEATHER BALLOONS
These lightweight, flexible balloons, commonly mistaken for UFOs, are routinely released around the globe to diagnose weather conditions. They can reach altitudes of 25 miles (40km) or more.

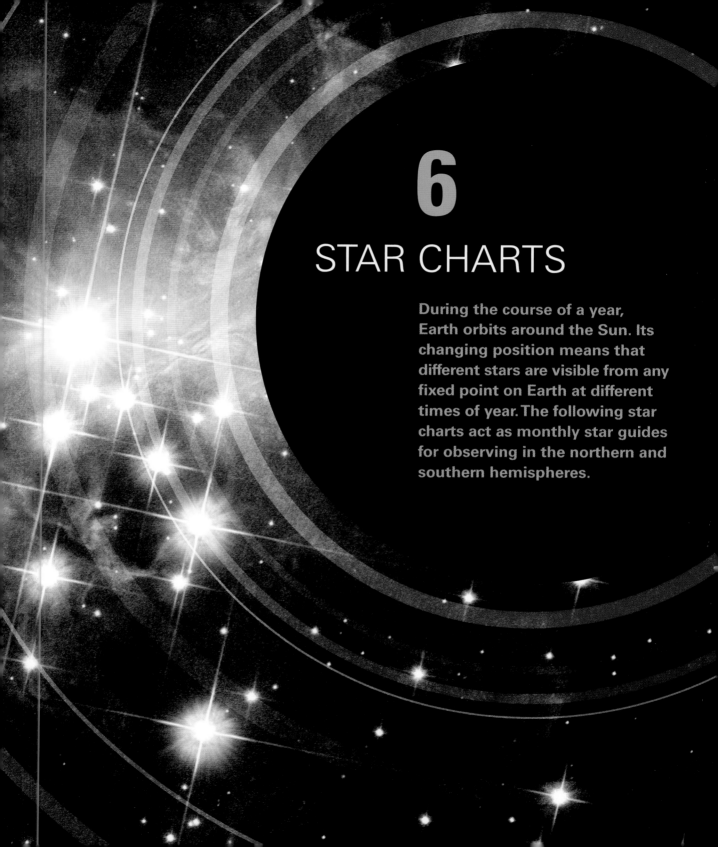

6

STAR CHARTS

During the course of a year, Earth orbits around the Sun. Its changing position means that different stars are visible from any fixed point on Earth at different times of year. The following star charts act as monthly star guides for observing in the northern and southern hemispheres.

USING THE STAR CHARTS

Star charts are excellent for naked-eye observing. They are great for planning a night's observing, as well as useful for starhopping from target to target.

Monthly sky maps

This section contains sky charts representing the night sky during each month of the year for observers at both northern and southern latitudes. They show some of the more prominent deep-sky objects visible, such as star clusters and nebulae. This page shows you how to use the charts and get the most from the information they provide.

THE NAKED EYE
Star charts are useful for observing the constellations with the naked eye and locating the brighter deep-sky objects.

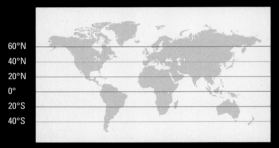

1 The first thing to do when using these star charts is to find the latitude nearest to your observing location. Look at the world map (left) and find the colored line of latitude closest to your location. (Note that a 10° difference in latitude will have only a small effect on the stars you can actually see.)

60°N
40°N
20°N
0°
20°S
40°S

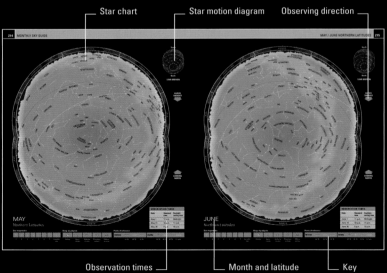

Star chart — Star motion diagram — Observing direction

MAY
Northern Latitudes

JUNE
Northern Latitudes

Observation times — Month and latitude — Key

2 The charts for a particular month show the sky as it appears at 10 p.m. in mid-month—that is, at 10 p.m. standard time for a particular time zone. When daylight-saving time (DST or summer time) is in use, they show the sky as it appears one hour later. The sky will look the same at 11 p.m. at the start of the month, and at 9 p.m. at the end of the month, as it does at 10 p.m. in mid-month. If you wish to look at the sky at a different time, you will need to use charts shown under a different month: for every two hours of time before or after 10 p.m., go one month backward or forward. So if you want to look at the sky at midnight on January 15, turn to the chart for February.

WHAT THE STAR CHARTS SHOW

FEATURE	DESCRIPTION
The Milky Way	A pale band across the chart represents the Milky Way. The charts can help you find good months to observe the Milky Way. Look for months when the band crosses right across the charts—for example, February for southern latitudes and September for northern latitudes.
The constellations	The constellations are marked on each chart with lines connecting their brighter stars. To find out more about the constellations and their deep-sky objects, refer to Chapter 4.
Nebulae	The locations of several of the night sky's finest nebulae are shown on the charts.
Stars	The charts show the stars in the night sky down to a magnitude of 5, well within naked-eye visibility, making the charts handy for starhopping with the naked eye or a pair of binoculars. The brightness of the stars is shown by the size of the star representing them. A key can be found on each chart for reference purposes. Variable stars are also plotted on these charts.
The zenith	The zenith (the point right above the observer's head) is marked for the different latitudes on each chart. The zenith for each latitude is coordinated so it matches the same color as the horizon indicator for each latitude. Refer to the key and the latitude map (above left) to check which zenith is appropriate for your observing location.
Galaxies	A selection of the brightest galaxies in the night sky is shown on the charts; each galaxy is represented by a small galaxy symbol.
Open and globular star clusters	Prominent open and globular clusters are marked on the charts along with their Messier or NGC numbers. They make good targets for small telescopes and the brighter examples can also be fine sights through a pair of binoculars.
The ecliptic	The ecliptic, the Sun's path across the sky, is marked as a bright yellow line on the star charts. Although planets are not plotted, you will find that they stay relatively close to the ecliptic.

3 Turn to the chart that shows the month in which you are observing and that corresponds to your location (either northern or southern latitudes). Facing north (a compass will help locate north), hold the book in front of you, perpendicular to the ground, with the NORTH label (at the bottom edge of the chart) pointing to the ground.

4 Locate the colored line on the chart that corresponds to your latitude. This will form a circle containing the area of the sky you can observe from this latitude.

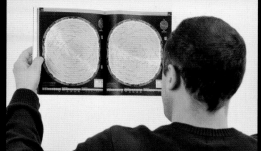

ORIENTING THE CHARTS
To view the northern part of the sky, turn to face north and hold the chart so that the "LOOKING NORTH" arrow points up. The stars' positions are correct for the times on each chart, so take this into account when observing at different times. To view the south, turn both yourself and the chart around to correctly orient the chart to the sky.

5 The curve of this circle closest to the bottom edge of the chart represents the northern horizon, directly in front of you. The opposite curve of the circle is the southern horizon, directly behind you. You will be unable to see any stars beyond this circle. The chart also shows a cross, corresponding in color to your line of latitude, in the center of the circle. This represents the zenith, the highest point in the sky directly above you.

6 The yellow line across the center of the chart represents the ecliptic. The Moon and the planets stay relatively close to the ecliptic when they are visible in the night sky.

7 To observe the sky to the south of your location, face the opposite direction and turn the book so that the SOUTH label is nearest to the ground. The bottom line of the circle is now the southern horizon, and the top line is the northern horizon.

8 Each chart also contains a star motion diagram showing the direction that the stars will move in as the night progresses. Stars farther from the celestial poles will move farther than those close to it, eventually moving beyond the horizon. The stars closest to the poles will rotate around them without setting.

USING THE KEY
Stars of magnitude 5 and brighter are shown on the star charts. Icons are used to represent deep-sky objects of interest to the amateur astronomer.

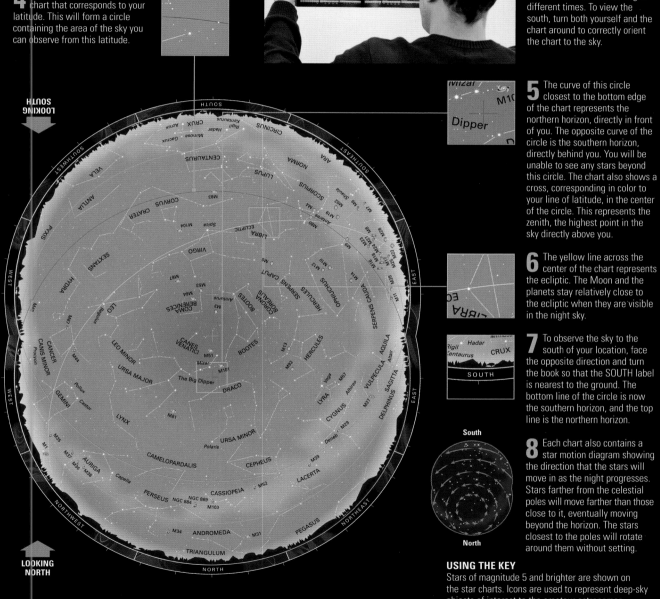

Star magnitudes							
-1	0	1	2	3	4	5	Variable star

Deep-sky objects				
Galaxy	Open cluster	Globular cluster	Planetary nebula	Diffuse nebula

Points of reference							
Horizons			Zeniths				
60°N	40°N	20°N	60°N	40°N	20°N	Ecliptic	

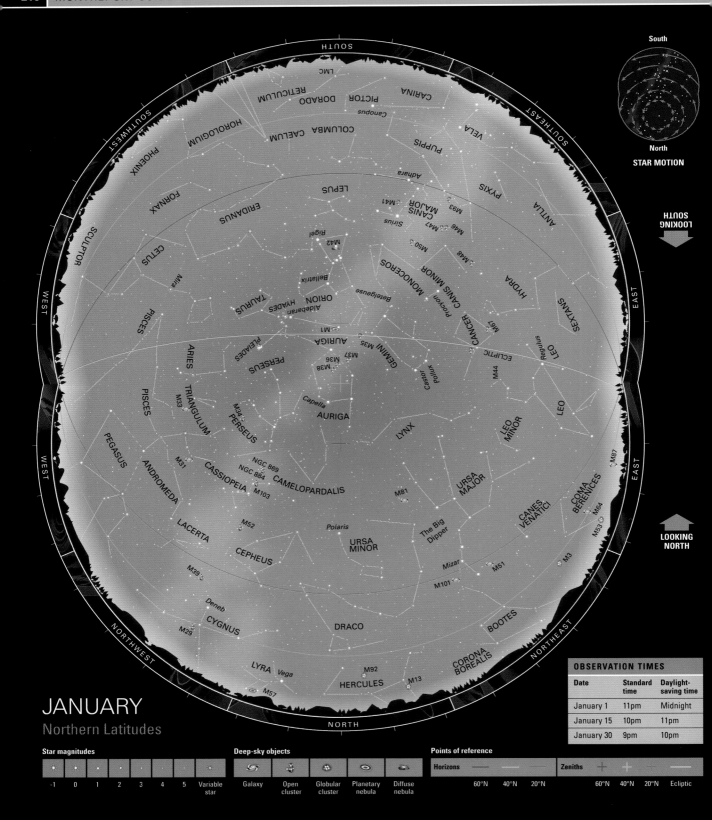

South

North

STAR MOTION

SOUTH LOOKING

LOOKING NORTH

JANUARY
Northern Latitudes

OBSERVATION TIMES		
Date	**Standard time**	**Daylight-saving time**
January 1	11pm	Midnight
January 15	10pm	11pm
January 30	9pm	10pm

Star magnitudes

-1	0	1	2	3	4	5	Variable star

Deep-sky objects

Galaxy | Open cluster | Globular cluster | Planetary nebula | Diffuse nebula

Points of reference

Horizons — 60°N 40°N 20°N

Zeniths — 60°N 40°N 20°N Ecliptic

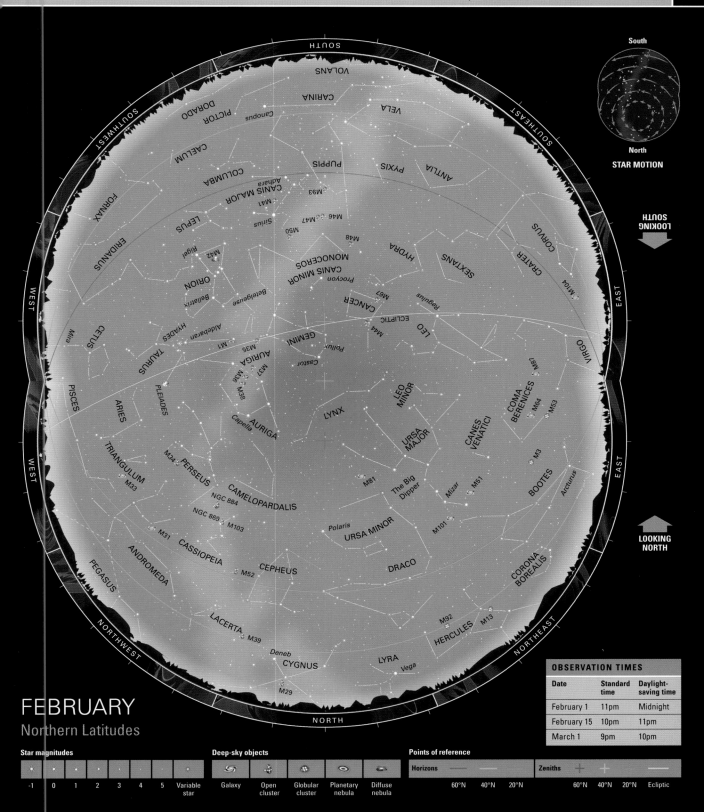

South

North

STAR MOTION

LOOKING SOUTH

LOOKING NORTH

FEBRUARY
Northern Latitudes

OBSERVATION TIMES

Date	Standard time	Daylight-saving time
February 1	11pm	Midnight
February 15	10pm	11pm
March 1	9pm	10pm

Star magnitudes

-1	0	1	2	3	4	5	Variable star

Deep-sky objects

Galaxy | Open cluster | Globular cluster | Planetary nebula | Diffuse nebula

Points of reference

Horizons ———

60°N 40°N 20°N

Zeniths

60°N 40°N 20°N Ecliptic

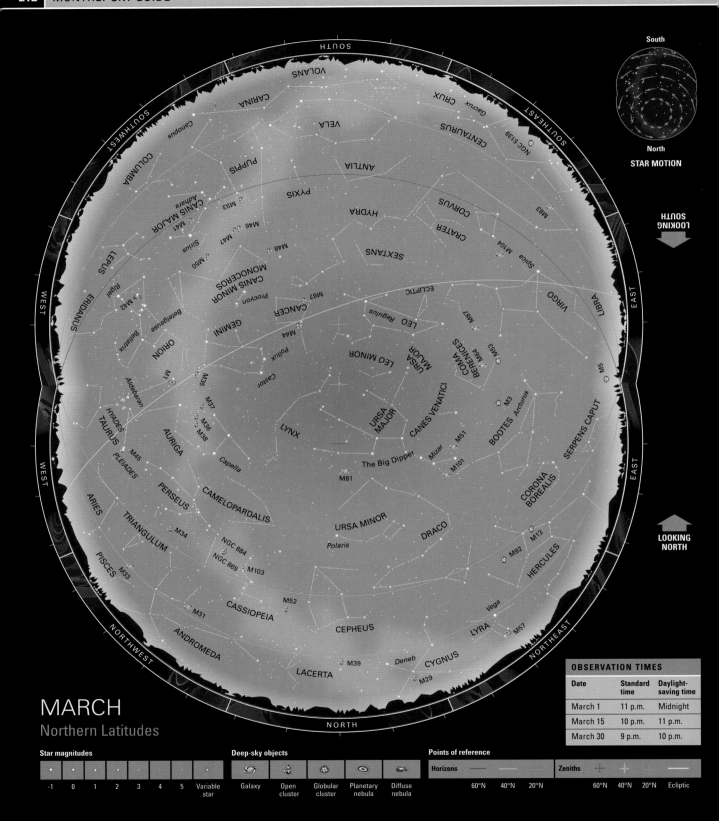

South

North

STAR MOTION

LOOKING
SOUTH

LOOKING
NORTH

MARCH
Northern Latitudes

OBSERVATION TIMES

Date	Standard time	Daylight-saving time
March 1	11 p.m.	Midnight
March 15	10 p.m.	11 p.m.
March 30	9 p.m.	10 p.m.

Star magnitudes

-1   0   1   2   3   4   5   Variable star

Deep-sky objects

Galaxy   Open cluster   Globular cluster   Planetary nebula   Diffuse nebula

Points of reference

Horizons   60°N   40°N   20°N   Zeniths   60°N   40°N   20°N   Ecliptic

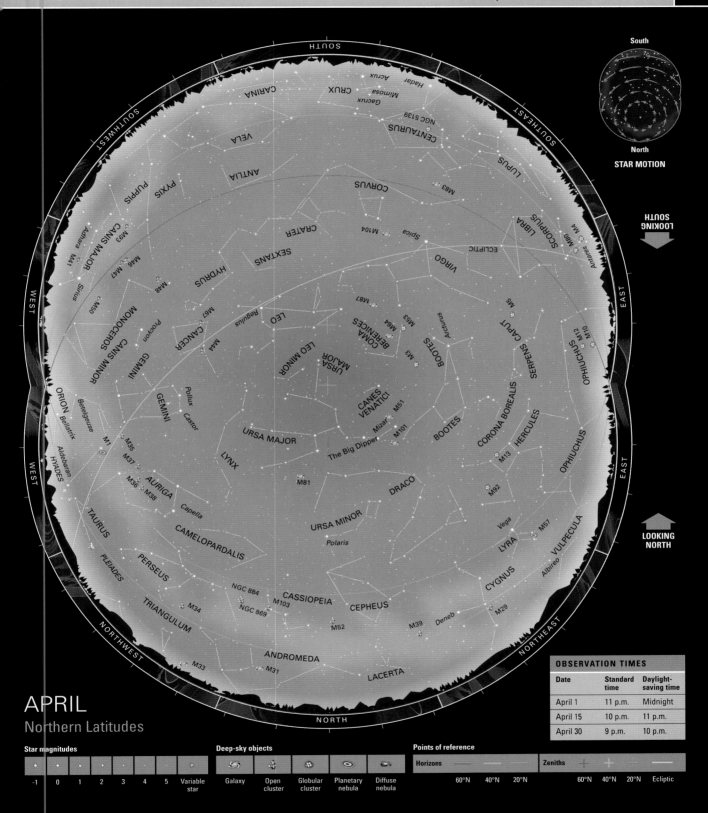

APRIL
Northern Latitudes

STAR MOTION

South

North

LOOKING SOUTH

LOOKING NORTH

OBSERVATION TIMES

Date	Standard time	Daylight-saving time
April 1	11 p.m.	Midnight
April 15	10 p.m.	11 p.m.
April 30	9 p.m.	10 p.m.

Star magnitudes

-1 0 1 2 3 4 5 Variable star

Deep-sky objects

Galaxy Open cluster Globular cluster Planetary nebula Diffuse nebula

Points of reference

Horizons   60°N 40°N 20°N

Zeniths   60°N 40°N 20°N Ecliptic

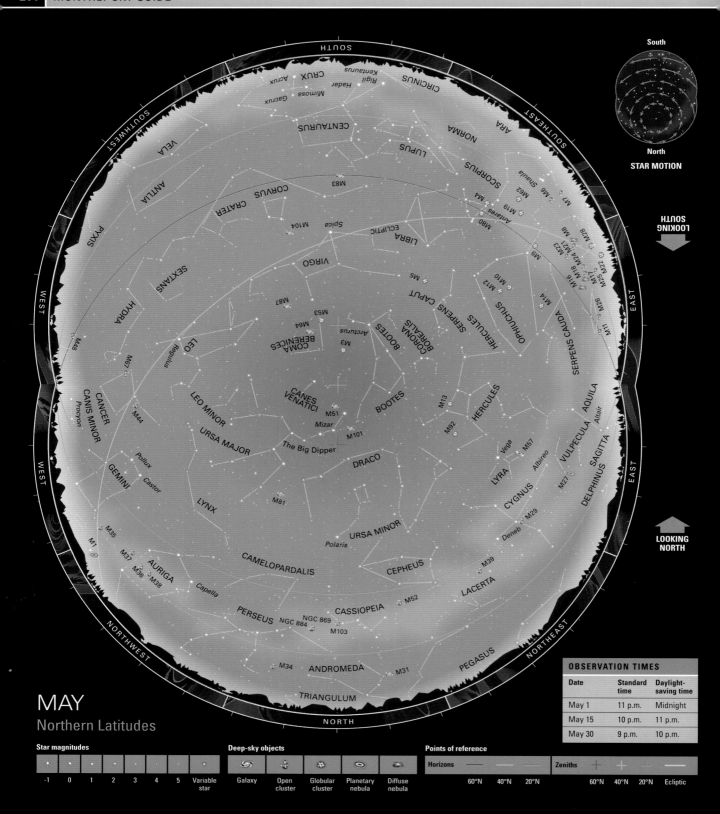

STAR MOTION

LOOKING SOUTH

LOOKING NORTH

MAY
Northern Latitudes

OBSERVATION TIMES

Date	Standard time	Daylight-saving time
May 1	11 p.m.	Midnight
May 15	10 p.m.	11 p.m.
May 30	9 p.m.	10 p.m.

Star magnitudes

-1	0	1	2	3	4	5	Variable star

Deep-sky objects

Galaxy   Open cluster   Globular cluster   Planetary nebula   Diffuse nebula

Points of reference

Horizons ——— 60°N 40°N 20°N

Zeniths + + + 60°N 40°N 20°N Ecliptic

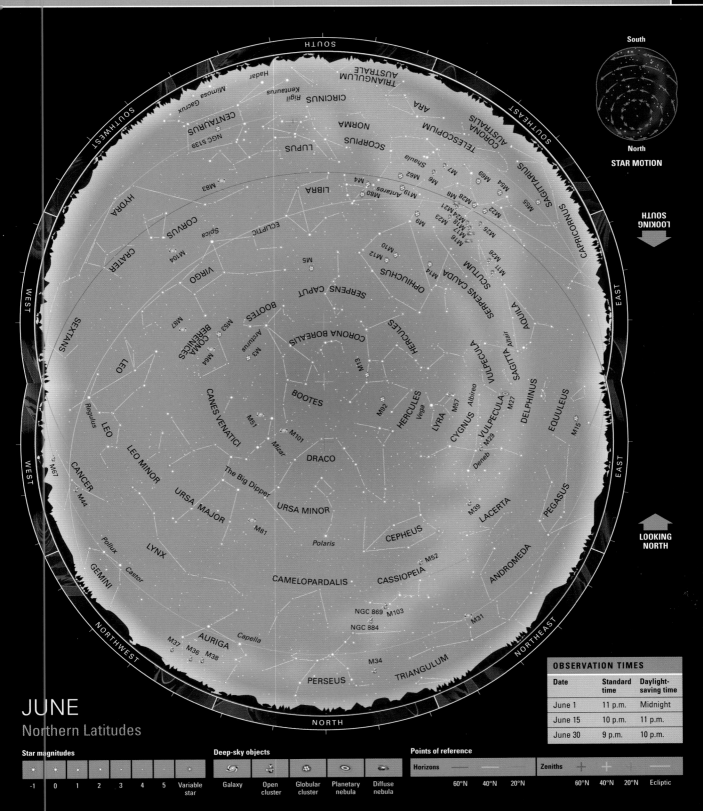

South

North

STAR MOTION

LOOKING
SOUTH

EAST

EAST

LOOKING
NORTH

JUNE
Northern Latitudes

OBSERVATION TIMES		
Date	Standard time	Daylight-saving time
June 1	11 p.m.	Midnight
June 15	10 p.m.	11 p.m.
June 30	9 p.m.	10 p.m.

Star magnitudes

| -1 | 0 | 1 | 2 | 3 | 4 | 5 | Variable star |

Deep-sky objects

Galaxy | Open cluster | Globular cluster | Planetary nebula | Diffuse nebula

Points of reference

Horizons ——— ——— ———   60°N 40°N 20°N

Zeniths + + —   60°N 40°N 20°N Ecliptic

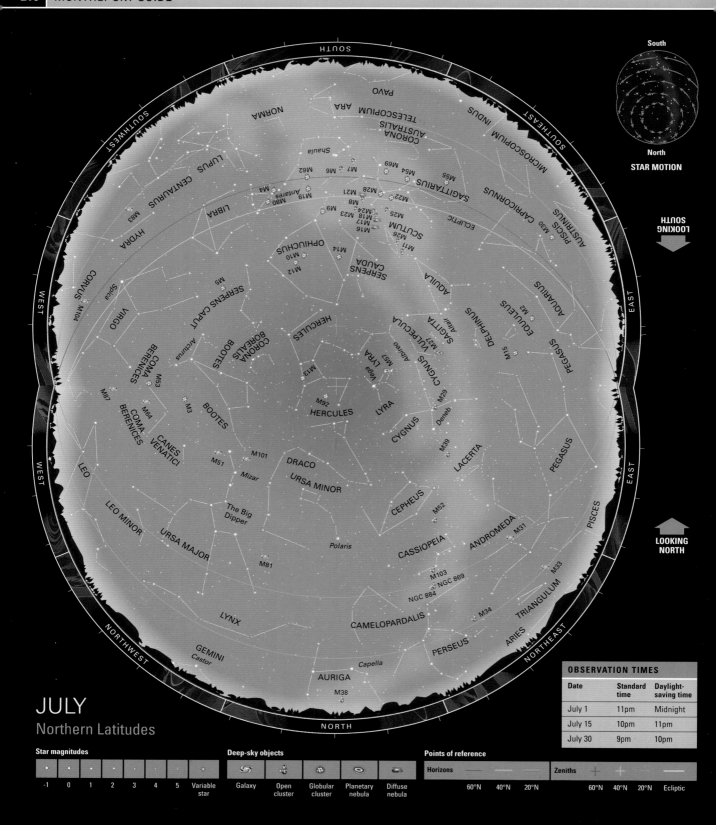

South
North
STAR MOTION

LOOKING
SOUTH

LOOKING
NORTH

JULY
Northern Latitudes

OBSERVATION TIMES

Date	Standard time	Daylight-saving time
July 1	11pm	Midnight
July 15	10pm	11pm
July 30	9pm	10pm

Star magnitudes

-1 0 1 2 3 4 5 Variable star

Deep-sky objects

Galaxy Open cluster Globular cluster Planetary nebula Diffuse nebula

Points of reference

Horizons ——— ——— ———   Zeniths + + —

60°N 40°N 20°N   60°N 40°N 20°N Ecliptic

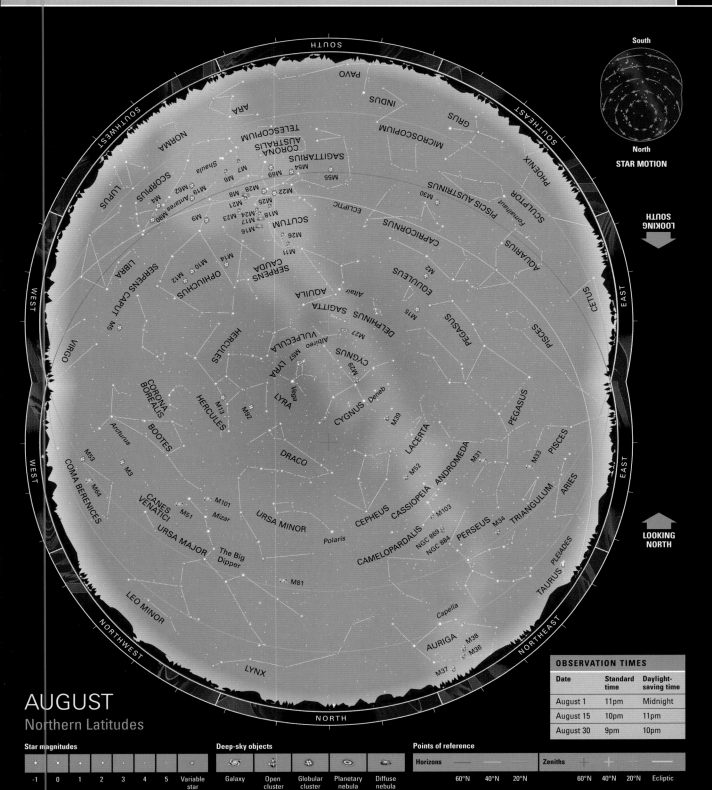

South

North

STAR MOTION

LOOKING SOUTH

LOOKING NORTH

AUGUST
Northern Latitudes

OBSERVATION TIMES

Date	Standard time	Daylight-saving time
August 1	11pm	Midnight
August 15	10pm	11pm
August 30	9pm	10pm

Star magnitudes

| -1 | 0 | 1 | 2 | 3 | 4 | 5 | Variable star |

Deep-sky objects

Galaxy   Open cluster   Globular cluster   Planetary nebula   Diffuse nebula

Points of reference

Horizons   60°N   40°N   20°N

Zeniths   60°N   40°N   20°N   Ecliptic

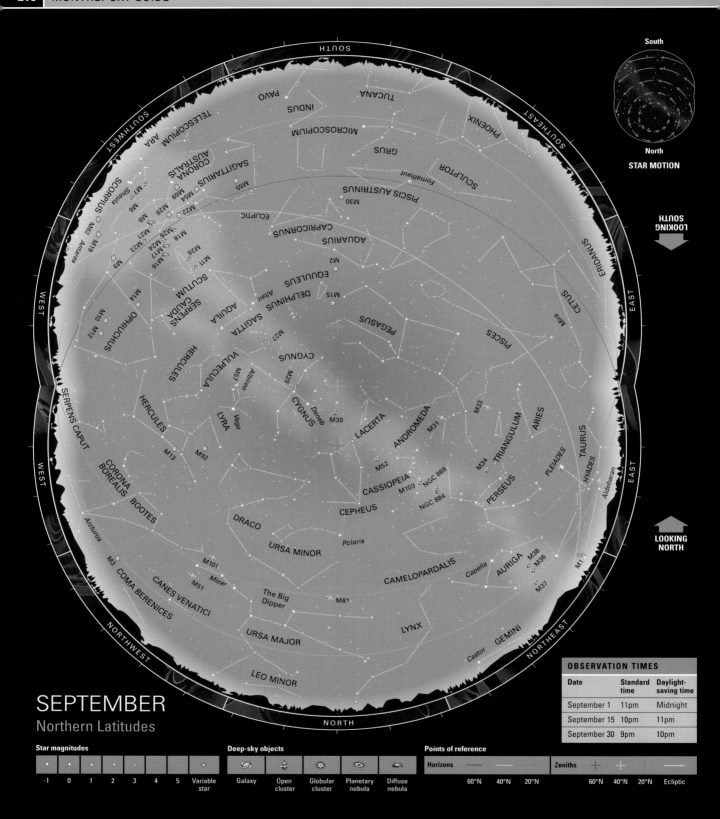

South

North

STAR MOTION

LOOKING SOUTH

LOOKING NORTH

SEPTEMBER
Northern Latitudes

OBSERVATION TIMES

Date	Standard time	Daylight-saving time
September 1	11pm	Midnight
September 15	10pm	11pm
September 30	9pm	10pm

Star magnitudes

-1	0	1	2	3	4	5	Variable star

Deep-sky objects

Galaxy Open cluster Globular cluster Planetary nebula Diffuse nebula

Points of reference

Horizons — 60°N 40°N 20°N

Zeniths — 60°N 40°N 20°N Ecliptic

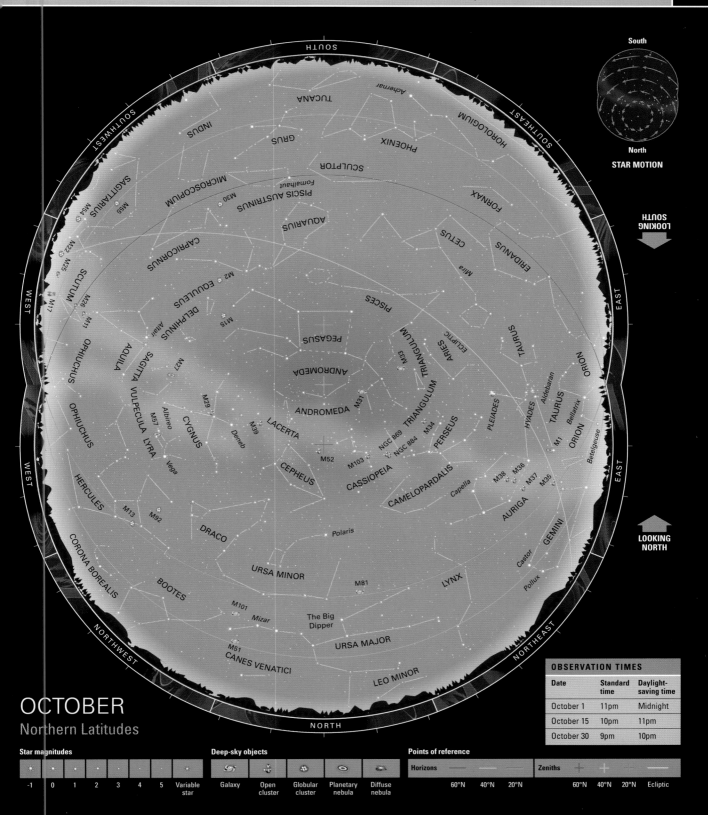

STAR MOTION

South

North

LOOKING SOUTH

LOOKING NORTH

OCTOBER
Northern Latitudes

OBSERVATION TIMES		
Date	Standard time	Daylight-saving time
October 1	11pm	Midnight
October 15	10pm	11pm
October 30	9pm	10pm

Star magnitudes

-1   0   1   2   3   4   5   Variable star

Deep-sky objects

Galaxy   Open cluster   Globular cluster   Planetary nebula   Diffuse nebula

Points of reference

Horizons   60°N   40°N   20°N

Zeniths   60°N   40°N   20°N   Ecliptic

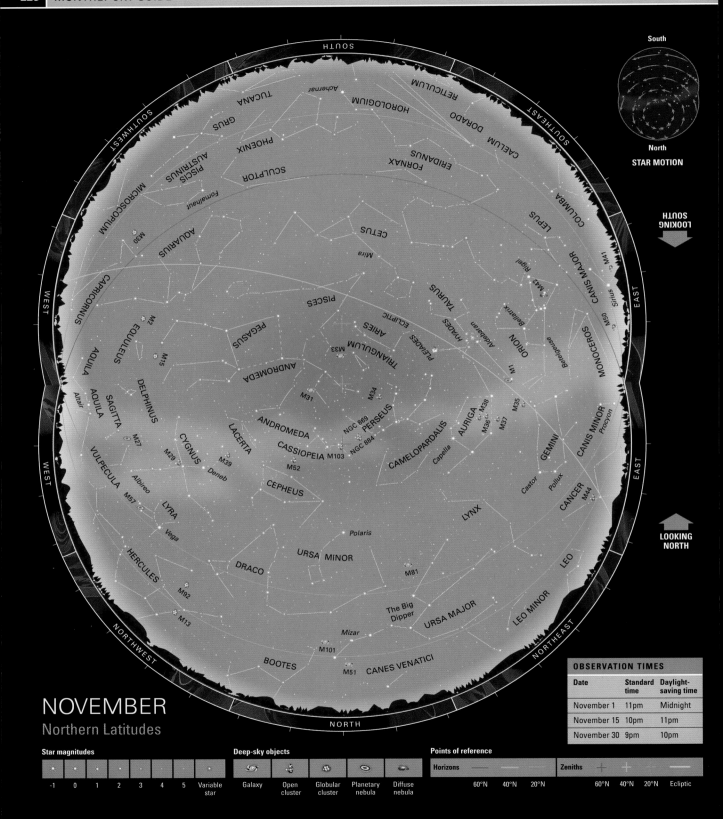

South

North

STAR MOTION

LOOKING SOUTH

LOOKING NORTH

NOVEMBER
Northern Latitudes

OBSERVATION TIMES

Date	Standard time	Daylight-saving time
November 1	11pm	Midnight
November 15	10pm	11pm
November 30	9pm	10pm

Star magnitudes

★	★	★	·	·	·	·	·
-1	0	1	2	3	4	5	Variable star

Deep-sky objects

🌀	✦	⬡	◉	☁
Galaxy	Open cluster	Globular cluster	Planetary nebula	Diffuse nebula

Points of reference

Horizons ———   Zeniths + + —

60°N   40°N   20°N   60°N   40°N   20°N   Ecliptic

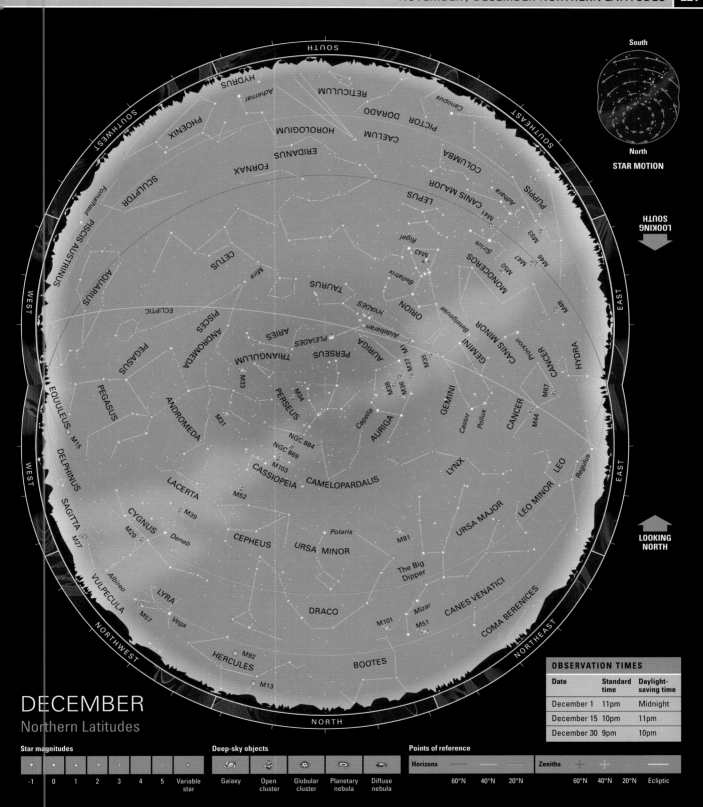

South

North

STAR MOTION

LOOKING SOUTH

LOOKING NORTH

DECEMBER
Northern Latitudes

OBSERVATION TIMES

Date	Standard time	Daylight-saving time
December 1	11pm	Midnight
December 15	10pm	11pm
December 30	9pm	10pm

Star magnitudes

-1	0	1	2	3	4	5	Variable star

Deep-sky objects

Galaxy	Open cluster	Globular cluster	Planetary nebula	Diffuse nebula

Points of reference

Horizons   60°N 40°N 20°N

Zeniths   60°N 40°N 20°N Ecliptic

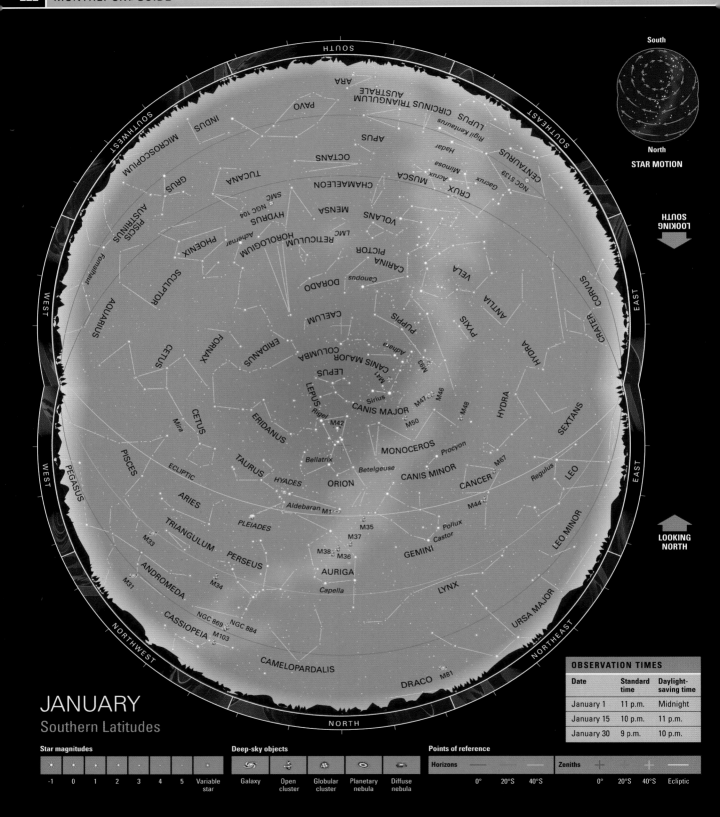

South

North

STAR MOTION

LOOKING SOUTH

LOOKING NORTH

JANUARY
Southern Latitudes

OBSERVATION TIMES

Date	Standard time	Daylight-saving time
January 1	11 p.m.	Midnight
January 15	10 p.m.	11 p.m.
January 30	9 p.m.	10 p.m.

Star magnitudes

-1	0	1	2	3	4	5	Variable star

Deep-sky objects

Galaxy	Open cluster	Globular cluster	Planetary nebula	Diffuse nebula

Points of reference

Horizons			Zeniths			
0°	20°S	40°S	0°	20°S	40°S	Ecliptic

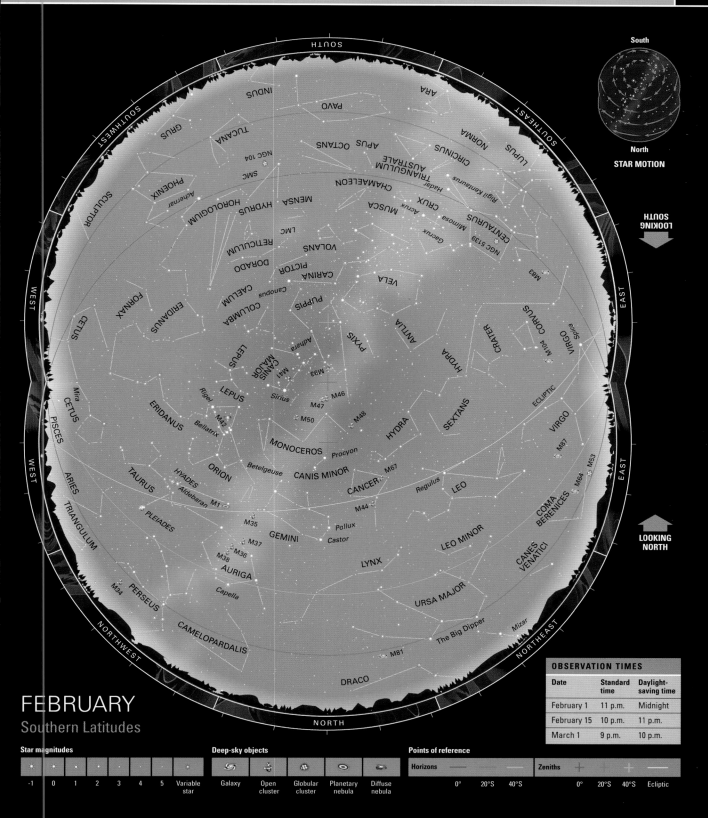

South

North

STAR MOTION

LOOKING SOUTH

LOOKING NORTH

FEBRUARY
Southern Latitudes

OBSERVATION TIMES

Date	Standard time	Daylight-saving time
February 1	11 p.m.	Midnight
February 15	10 p.m.	11 p.m.
March 1	9 p.m.	10 p.m.

Star magnitudes

-1	0	1	2	3	4	5	Variable star

Deep-sky objects

Galaxy | Open cluster | Globular cluster | Planetary nebula | Diffuse nebula

Points of reference

Horizons ——   0°   20°S   40°S

Zeniths   0°   20°S   40°S   Ecliptic

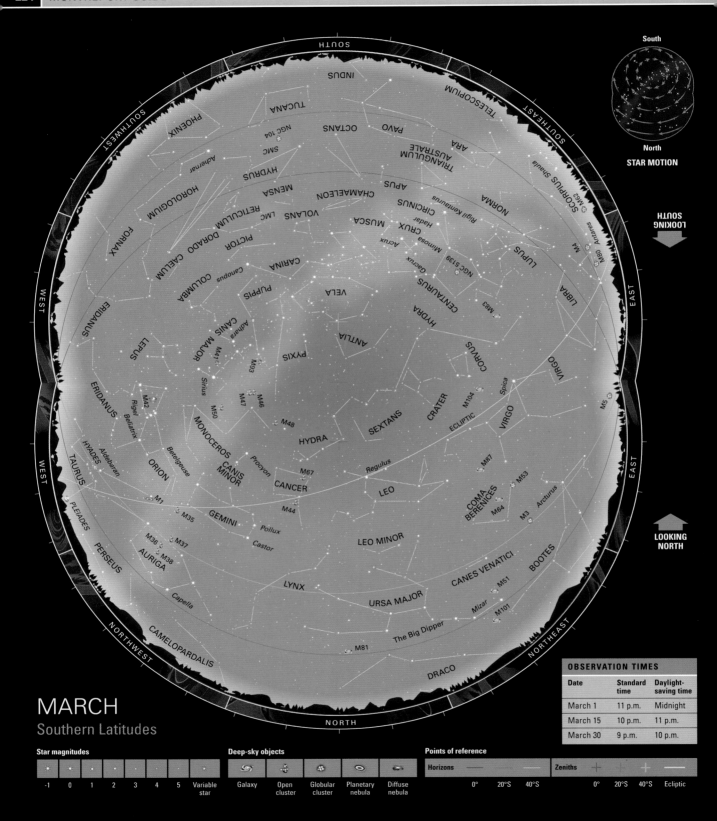

STAR MOTION

South

North

LOOKING SOUTH

LOOKING NORTH

MARCH
Southern Latitudes

Star magnitudes

-1	0	1	2	3	4	5	Variable star

Deep-sky objects

Galaxy | Open cluster | Globular cluster | Planetary nebula | Diffuse nebula

Points of reference

Horizons ——— Zeniths

0° 20°S 40°S   0° 20°S 40°S Ecliptic

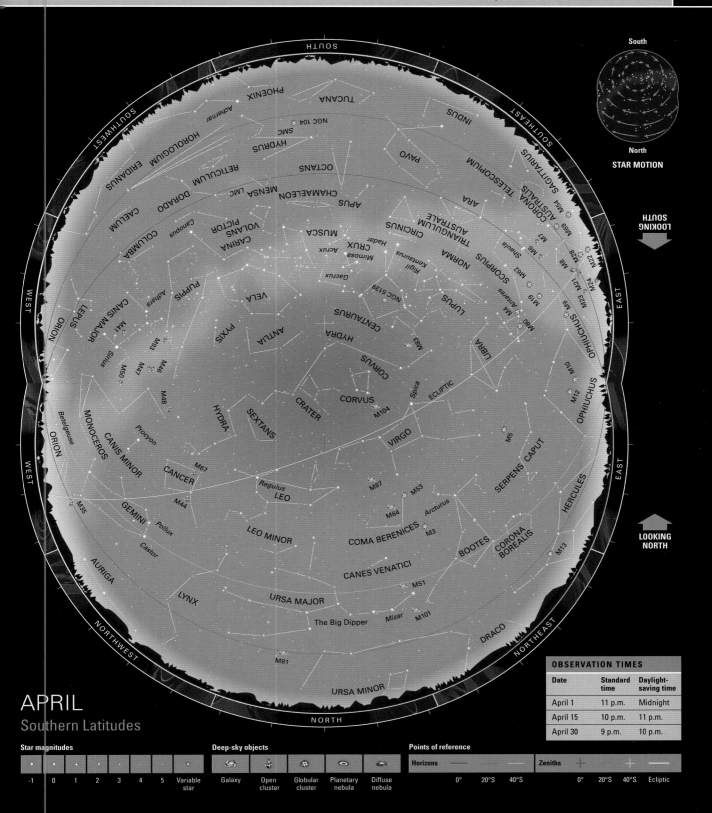

STAR MOTION

South

North

LOOKING
SOUTH

LOOKING
NORTH

APRIL
Southern Latitudes

OBSERVATION TIMES

Date	Standard time	Daylight-saving time
April 1	11 p.m.	Midnight
April 15	10 p.m.	11 p.m.
April 30	9 p.m.	10 p.m.

Star magnitudes

-1	0	1	2	3	4	5	Variable star

Deep-sky objects

Galaxy	Open cluster	Globular cluster	Planetary nebula	Diffuse nebula

Points of reference

Horizons — 0° 20°S 40°S

Zeniths + 0° 20°S 40°S Ecliptic

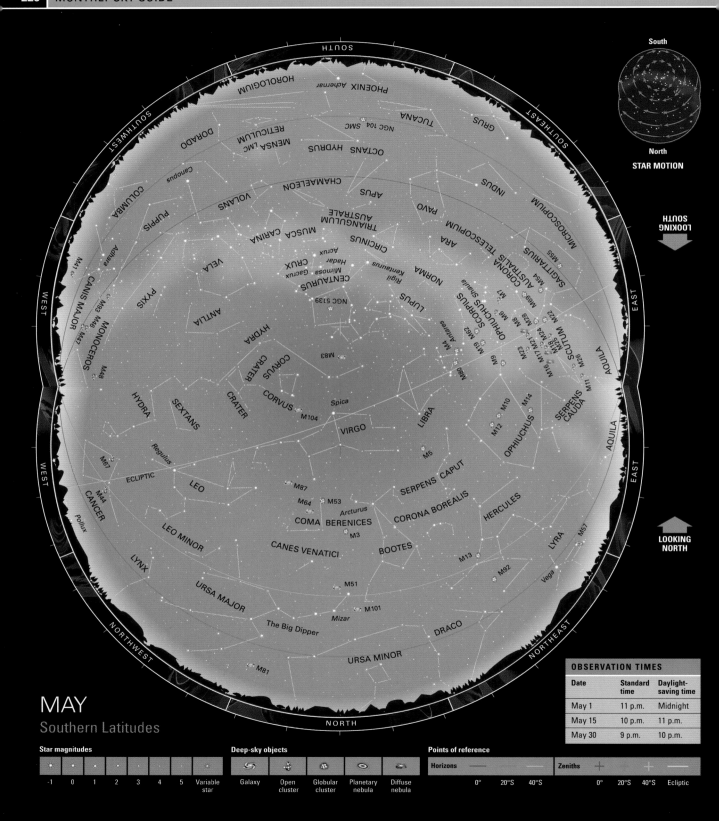

South

North

STAR MOTION

LOOKING SOUTH

LOOKING NORTH

MAY
Southern Latitudes

OBSERVATION TIMES

Date	Standard time	Daylight-saving time
May 1	11 p.m.	Midnight
May 15	10 p.m.	11 p.m.
May 30	9 p.m.	10 p.m.

Star magnitudes

| -1 | 0 | 1 | 2 | 3 | 4 | 5 | Variable star |

Deep-sky objects

Galaxy | Open cluster | Globular cluster | Planetary nebula | Diffuse nebula

Points of reference

Horizons: 0° 20°S 40°S

Zeniths: 0° 20°S 40°S Ecliptic

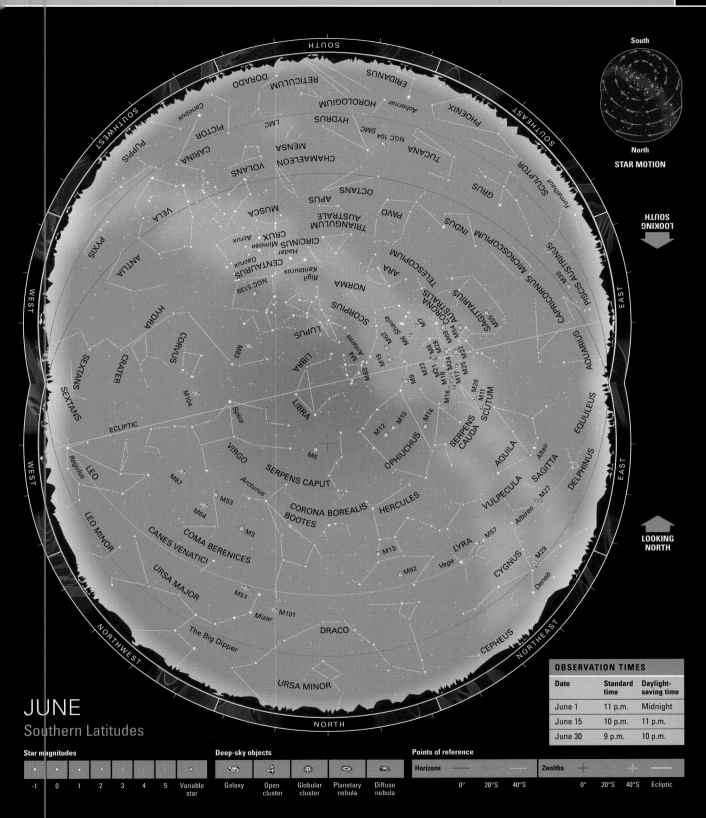

STAR MOTION

LOOKING SOUTH

LOOKING NORTH

JUNE
Southern Latitudes

Star magnitudes

| -1 | 0 | 1 | 2 | 3 | 4 | 5 | Variable star |

Deep-sky objects

| Galaxy | Open cluster | Globular cluster | Planetary nebula | Diffuse nebula |

Points of reference

| Horizons | | Zeniths | | |
| 0° | 20°S | 40°S | 0° | 20°S | 40°S | Ecliptic |

OBSERVATION TIMES

Date	Standard time	Daylight-saving time
June 1	11 p.m.	Midnight
June 15	10 p.m.	11 p.m.
June 30	9 p.m.	10 p.m.

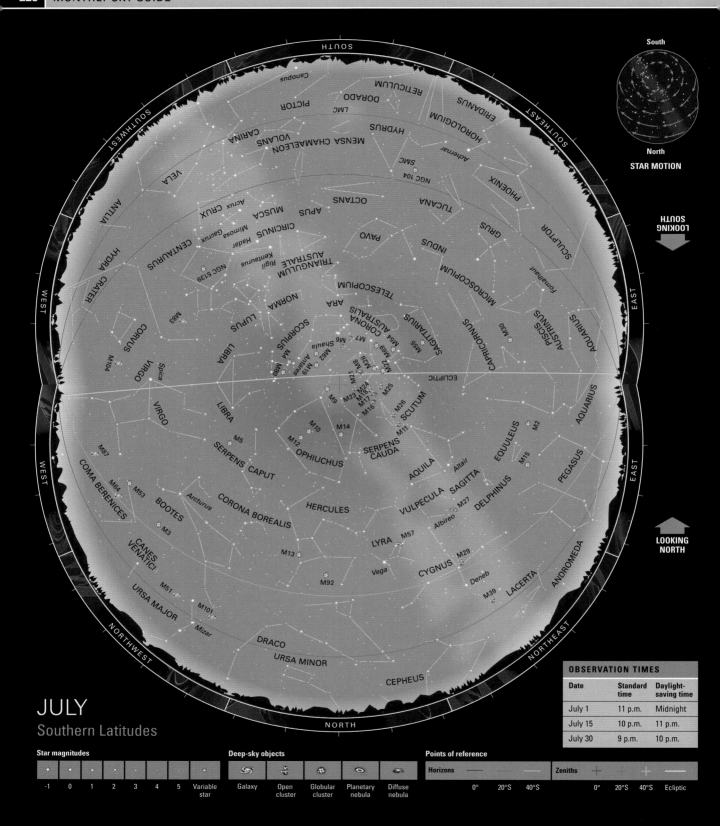

South

North

STAR MOTION

LOOKING SOUTH

LOOKING NORTH

JULY
Southern Latitudes

OBSERVATION TIMES

Date	Standard time	Daylight-saving time
July 1	11 p.m.	Midnight
July 15	10 p.m.	11 p.m.
July 30	9 p.m.	10 p.m.

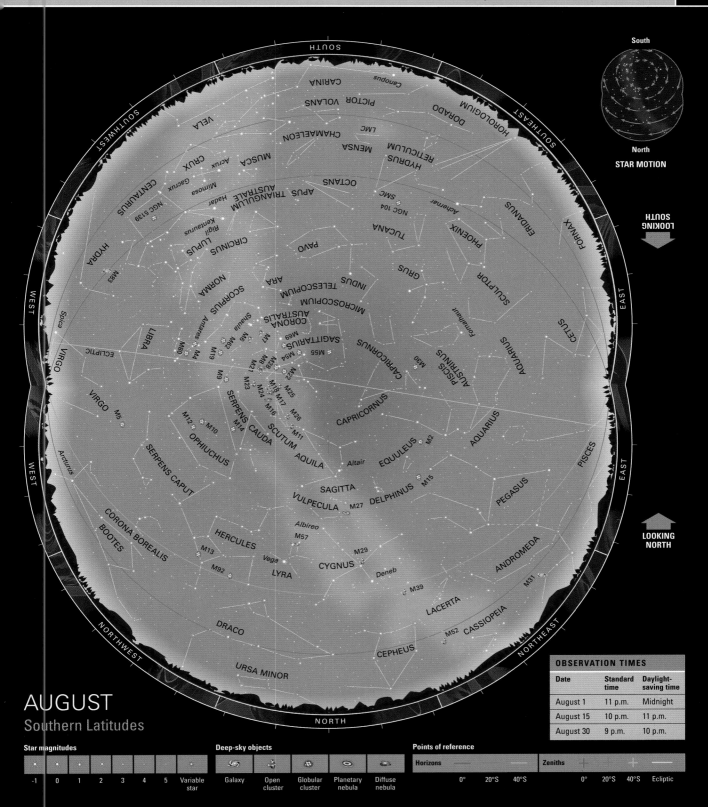

South

North

STAR MOTION

LOOKING SOUTH

LOOKING NORTH

AUGUST
Southern Latitudes

OBSERVATION TIMES

Date	Standard time	Daylight-saving time
August 1	11 p.m.	Midnight
August 15	10 p.m.	11 p.m.
August 30	9 p.m.	10 p.m.

Star magnitudes

-1   0   1   2   3   4   5   Variable star

Deep-sky objects

Galaxy   Open cluster   Globular cluster   Planetary nebula   Diffuse nebula

Points of reference

Horizons   0°   20°S   40°S   Zeniths   0°   20°S   40°S   Ecliptic

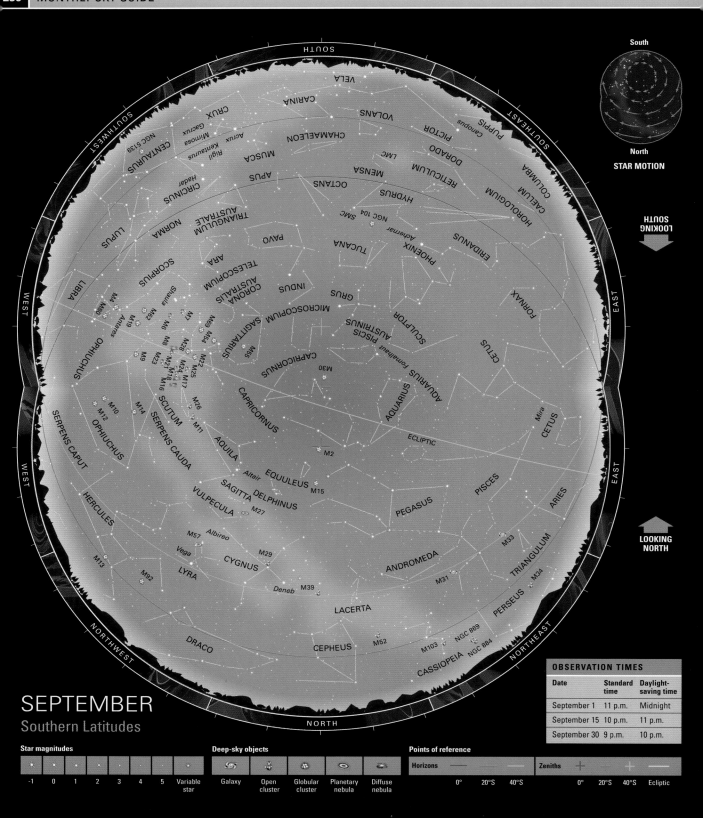

South

North

STAR MOTION

LOOKING SOUTH

LOOKING NORTH

SEPTEMBER
Southern Latitudes

Star magnitudes

| -1 | 0 | 1 | 2 | 3 | 4 | 5 | Variable star |

Deep-sky objects

Galaxy | Open cluster | Globular cluster | Planetary nebula | Diffuse nebula

Points of reference

Horizons   0°   20°S   40°S

Zeniths   0°   20°S   40°S   Ecliptic

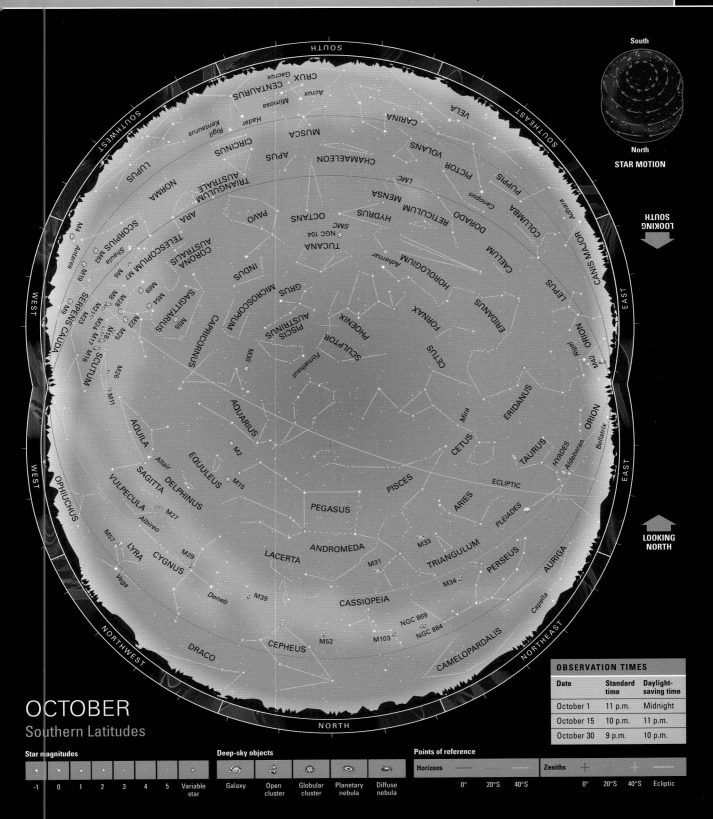

South

North

STAR MOTION

LOOKING SOUTH

LOOKING NORTH

OCTOBER
Southern Latitudes

OBSERVATION TIMES

Date	Standard time	Daylight-saving time
October 1	11 p.m.	Midnight
October 15	10 p.m.	11 p.m.
October 30	9 p.m.	10 p.m.

Star magnitudes

-1	0	1	2	3	4	5	Variable star

Deep-sky objects

Galaxy | Open cluster | Globular cluster | Planetary nebula | Diffuse nebula

Points of reference

Horizons   0° 20°S 40°S

Zeniths   0° 20°S 40°S Ecliptic

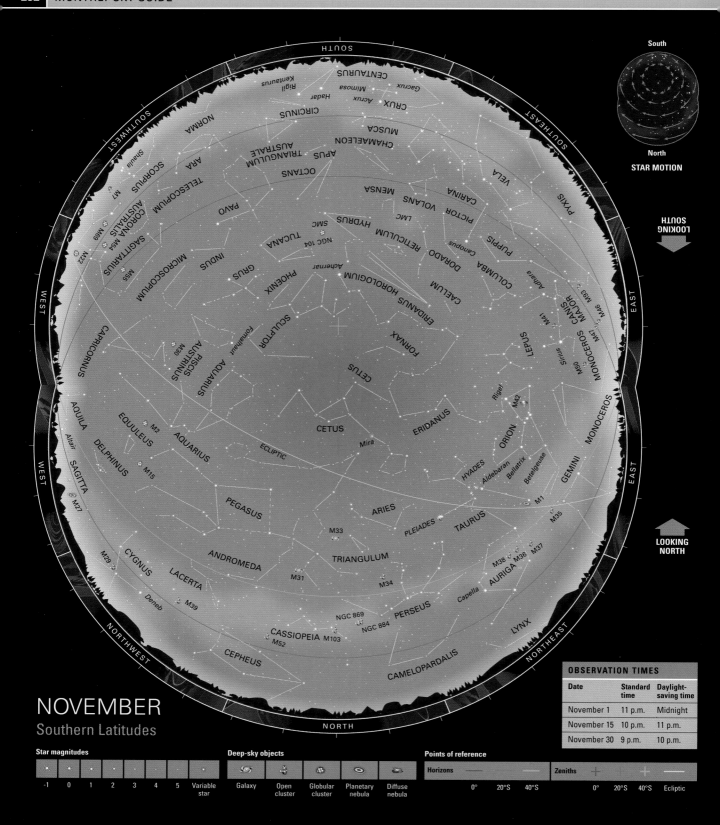

South

North

STAR MOTION

LOOKING SOUTH

LOOKING NORTH

NOVEMBER
Southern Latitudes

OBSERVATION TIMES

Date	Standard time	Daylight-saving time
November 1	11 p.m.	Midnight
November 15	10 p.m.	11 p.m.
November 30	9 p.m.	10 p.m.

Star magnitudes

-1	0	1	2	3	4	5	Variable star

Deep-sky objects

Galaxy · Open cluster · Globular cluster · Planetary nebula · Diffuse nebula

Points of reference

Horizons — 0° 20°S 40°S

Zeniths — 0° 20°S 40°S · Ecliptic

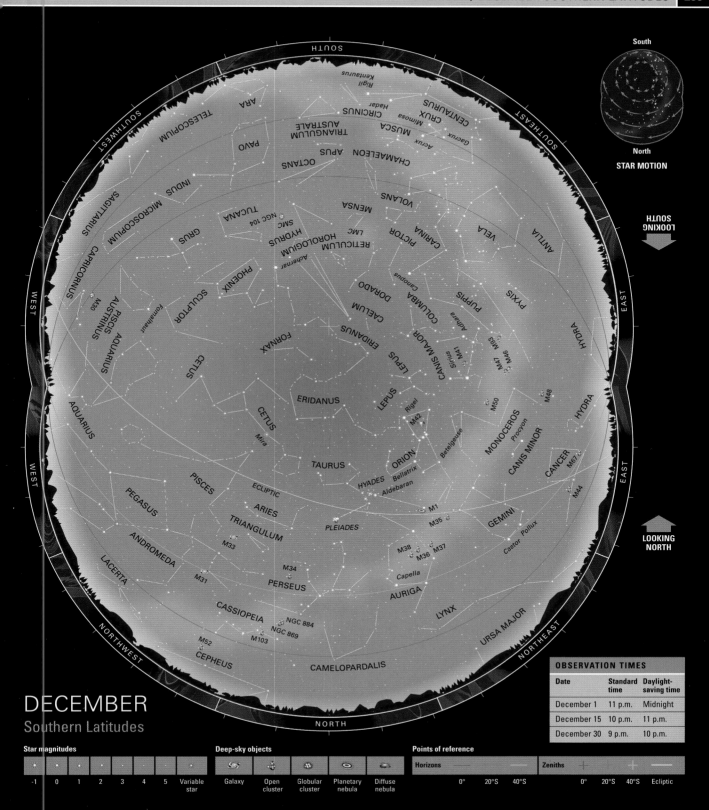

South

North

STAR MOTION

LOOKING
SOUTH

LOOKING
NORTH

DECEMBER
Southern Latitudes

OBSERVATION TIMES

Date	Standard time	Daylight-saving time
December 1	11 p.m.	Midnight
December 15	10 p.m.	11 p.m.
December 30	9 p.m.	10 p.m.

Star magnitudes

-1	0	1	2	3	4	5	Variable star

Deep-sky objects

Galaxy	Open cluster	Globular cluster	Planetary nebula	Diffuse nebula

Points of reference

Horizons — 0° — 20°S — 40°S   Zeniths + 0° + 20°S + 40°S — Ecliptic

7

REFERENCE

Important information for stargazers, including lists of deep-sky objects to observe, is presented here in a concise and easy-to-use form. There is also a handy glossary that clearly explains key astronomical terms.

REFERENCE TABLES

The tables listed here provide detailed information about the celestial bodies of the Solar System, such as their sizes, orbits, and physical properties. You will also find lists of notable stars and other deep-sky objects, as well as the Messier Catalog and a list of all 88 recognized constellations. You can use these tables as a "menu" to find interesting targets for a night's observation.

THE PLANETS

NAME	MERCURY	VENUS	EARTH	MARS	JUPITER	SATURN	URANUS	NEPTUNE
Equatorial diameter	3,029 miles (4,875km)	7,521 miles (12,104km)	7.926 miles (12,756km)	4,213 miles (6,780km)	88,846 miles (142,984km)	74,898 miles (120,536km)	31,763 miles (51,118km)	30,760 miles (49,532km)
Mass (in relation to Earth)	0.1	0.8	1	0.1	318	95	14.5	17.1
Volume (in relation to Earth)	0.1	0.9	1	0.2	1,321	763.6	63.1	57.7
Average surface/ cloud-top temperature	-292 to 806°F (-180 to 430°C) (surface)	867°F (464°C) (surface)	59°F (15°C) (surface)	-195 to 77°F (-125 to 25°C) (surface)	-160°F (-110°C) (cloud-top)	-220°F (-140°C) (cloud-top)	-353°F (-214°C) (cloud-top)	-364°F (-220°C) (cloud-top)
Gravity (in relation to Earth)	0.4	0.9	1	0.4	2.5	1.1	0.9	1.1
Number of moons	0	0	1	2	79	82	27+	13+
Mean distance from the Sun	36 million miles (57.9 million km)	67.2 million miles (108.2 million km)	93 million miles (149.6 million km)	141.6 million miles (227.9 million km)	483.6 million miles (778.3 million km)	888 million miles (1,430 million km)	1,784 million miles (2,871 million km)	2,794 million miles (4,497 million km)
Orbital period	88 days	224.7 days	365.3 days	687 days	11.9 years	29.5 years	84 years	164.8 years
Rotational period	59 days	243 days	23.9 hours	24.6 hours	9.9 hours	10.66 hours	17.24 hours	16.1 hours
Inclination of axis rotation	2.1°	177.3°	23.5°	25.2°	3.1°	26.7°	97.9°	29.6°
Apparent magnitude	-1.9 to 5.7	-4.6 to -3.8	n/a	-2.9 to -4.5	-2.9	-0.2 to 1.2	5.5	7.8

NOTABLE ASTEROIDS

NAME	GASPRA	ANNEFRANK	TOUTATIS	VESTA	MATHILDE	IDA	EROS
Average distance from the Sun	206 million miles (331 million km)	206 million miles (331 million km)	234 million miles (376 million km)	219 million miles (353 million km)	246 million miles (396 million km)	266 million miles (428 million km)	136 million miles (218 million km)
Orbital speed	44,470 mph (71,568 kph)	Unknown	37,334 mph (60,084 kph)	43,262 mph (69,624 kph)	40,270 mph (64,728 kph)	Unknown	54,492 mph (87,696 kph)
Orbital period	3.3 years	3.3 years	4.0 years	3.6 years	4.3 years	4.8 years	1.8 years
Rotation period	7 hours	Unknown	5.4 and 7.3 days (two axes)	5.3 hours	418 hours	4.6 hours	5.3 hours
Length	11.2 miles (18km)	3.7 miles (6km)	3.7 miles (4.3km)	348 miles (560km)	41 miles (66km)	37 miles (60km)	19.3 miles (31km)
Date discovered	July 30, 1916	March 23, 1942	January 4, 1989	March 27, 1807	November 12, 1885	September 29, 1884	August 13, 1898

ANNUAL METEOR SHOWERS

SHOWER	PEAK DATE (S)	DATE RANGE	MAX. NUMBER PER HOUR	NOTES
Quadrantids	Jan 3	Jan 1–6	90	Yellow and blue meteors traveling at medium speed.
Alpha Centaurids	Feb 8	Jan 28–Feb 21	20	A southern shower with some very bright, swift meteors.
Virginids	Apr 7–15	Mar 10–Apr 21	5	Slow, long trails with multiple radiants.
Lyrids	Apr 22	Apr 16–28	15	Fairly fast meteors from Comet Thatcher.
Eta Aquariids	May 6	Apr 21–May 24	60	A good southern shower with very fast, bright meteors from Comet Halley.
Arietids	Jun 7	May 22–Jun 30	55	A daytime shower, with some visible just before dawn.
June Boötids	Jun 28	Jun 27–30	Varies	A slow-speed variable shower with occasional strong outbursts. From Comet Pons–Winnecke.
Capricornids	Jul 5–20	Jun 10–Jul 30	5	Slow, yellow and blue bright meteors. Several peak dates and radiants.
Delta Aquariids	Jul 28–Aug 8	Jul 15–Aug 19	205	A southern shower with a double peak.
Piscis Austrinids	Jul 28	Jul 16–Aug 8	5	Southern-hemisphere shower of very slow meteors.
Alpha Capricornids	Aug 1	Jul 15–Aug 25	5	Produces slow fireballs visible for many seconds.
Perseids	Aug 12	Jul 23–Aug 22	90	Many bright meteors with trails, from Comet Swift–Tuttle.
Alpha Aurigids	Sep 1	Aug 25–Sep 7	7	Occasional bursts of over 100 per hour have been seen.
Giacobinids or Draconids	Oct 8	Oct 6–10	Variable	Slow meteors from Comet Giacobini–Zinner.
Orionids	Oct 21	Oct 5–30	25	Fast, with many trails. From Halley's Comet.
Taurids	Nov 4–12	Nov 1–25	10	Bright and slow meteors with two peaks from Comet Encke.
Leonids	Nov 17	Nov 14–21	Variable	Very fast meteors with trails, from Comet Temple–Tuttle.
Geminids	Dec 14	Dec 6–18	100	Medium-speed, bright meteors from asteroid Phaethon (3200).
Ursids	Dec 22	Dec 17–25	10	Slow meteors from Comet Tuttle.

LEONID METEOR SHOWER (LONG-EXPOSURE IMAGE)

BRIGHTEST STARS VIEWED FROM THE EARTH

RANK	APPARENT MAGNITUDE	BAYER DESIGNATION	NAME	DISTANCE FROM EARTH (LIGHT-YEARS)
1	-1.4	Alpha (α) Canis Majoris	Sirius	8.6
2	-0.7	Alpha (α) Carinae	Canopus	310
3	-0.3	Alpha (α) Centauri	Rigil Kentaurus	4.4
4	-0.04	Alpha (α) Boötis	Arcturus	75
5	0.0	Alpha (α) Lyrae	Vega	26
6	0.1	Alpha (α) Aurigae	Capella	42
7	0.2	Beta (β) Orionis	Rigel	770
8	0.4	Alpha (α) Canis Minoris	Procyon	11.6
9	0.5	Alpha (α) Eridani	Achernar	140
10	0.0–1.3	Alpha (α) Orionis	Betelgeuse	650
11	0.6	Beta (β) Centauri	Hadar	525
12	0.8	Alpha (α) Aquilae	Altair	16.8
13	0.8–1.0	Alpha (α) Tauri	Aldebaran	65
14	0.8	Alpha (α) Crucis	Acrux	320
15	1.0	Alpha (α) Virginis	Spica	220
16	1.0	Alpha (α) Scorpii	Antares	600
17	1.2	Beta (β) Geminorum	Pollux	34
18	1.2	Alpha (α) Piscis Austrini	Fomalhaut	25
19	1.3	Alpha (α) Cygni	Deneb	1,500
20	1.3	Beta (β) Crucis	Mimosa	350

CLOSEST STARS TO THE EARTH

RANK	STAR	DISTANCE FROM EARTH (LY)	CONSTELLATION	APPARENT MAGNITUDE
1	Alpha (α) Centauri System:		Centaurus	
	Proxima Centauri	4.2		11.1
	Alpha (α) Centauri	4.4		-0.3
2	Barnard's Star	6.0	Ophiuchus	9.6
3	Wolf 359	7.8	Leo	13.4
4	Lalande 21185	8.3	Ursa Major	7.5
5	Sirius	8.6	Canis Major	-1.4
6	Luyten 726–8	8.7	Cetus	12.5
7	Ross 154	9.7	Sagittarius	10.4
8	Ross 248	10.3	Andromeda	12.3
9	Epsilon (ε) Eridani	10.5	Eridanus	3.7
10	Lacaille 9352	10.7	Piscis Austrinus	7.3
11	Ross 128	10.9	Virgo	11.1
12	EZ Aquarii	11.3	Aquarius	13.3
13	Procyon	11.4	Canis Minor	0.4
14	Bessel's star	11.4	Cygnus	5.2
15	Struve 2398	11.5	Draco	8.9
16	Groombridge 34	11.6	Andromeda	8.1
17	Epsilon (ε) Indi	11.8	Indus	4.7
18	DX Cancri	11.8	Cancer	14.8
19	Tau (τ) Ceti	11.9	Cetus	3.5
20	GJ 1061	12.1	Horologium	13.1

GLOBULAR STAR CLUSTER M15

STAR CLUSTERS TO OBSERVE

CLUSTER NAME	CLUSTER TYPE	CONSTELLATION	APPARENT MAGNITUDE	DISTANCE FROM EARTH (LIGHT-YEARS)
47 Tucanae	Globular	Tucana	4.0	13,400
Beehive Cluster	Open	Cancer	3.7	577
Butterfly Cluster	Open	Scorpius	4.2	2,000
Hyades	Open	Taurus	4.2	150
Jewel Box	Open	Crux	7.1	8,150
M4	Globular	Scorpius	6.0	6,800
M12	Globular	Ophiuchus	6.6	16,000–18,000
M14	Globular	Ophiuchus	6.4	23,000–30,000
M15	Globular	Pegasus	6.2	33,600
M52	Open	Cassiopeia	7.3	5,000
M68	Globular	Hydra	7.5	33,000–44,000
M93	Open	Puppis	6.2	3,600
M107	Globular	Ophiuchus	8.9	20,900
NGC 3201	Globular	Vela	8.2	15,000
NGC 4833	Globular	Musca	5.3	17,000
Omega (ω) Centauri	Globular	Centaurus	4.2	17,000
Pleiades	Open	Taurus	1.5	380

MULTIPLE STARS TO OBSERVE

NAME	CONSTELLATION	NUMBER OF STARS	APPARENT MAGNITUDE	DISTANCE FROM EARTH (LIGHT-YEARS)
Castor	Gemini	6	1.9	50
Sigma (σ) Orionis	Orion	5	3.8	1,150
Alcyone	Taurus	4	2.9	368
Algol	Perseus	4	2.1–3.4	93
Almach	Andromeda	4	2.3	355
Epsilon (ε) Lyrae	Lyra	4	4.7	160
Mizar & Alcor	Ursa Major	4	2.3	81
Theta (θ) Orionis	Orion	4	4.7	1,800
Albireo	Cygnus	3	3.2	385
Beta (β) Monocerotis	Monoceros	3	3.7	700
Omicron (o) Eridani	Eridanus	3	9.5	16
Rigel	Orion	3	0.2	770
15 Monocerotis	Monoceros	2	4.7	1,020
Epsilon (ε) Aurigae	Auriga	2	3.1	2,040
Izar	Boötes	2	2.4	210
M40	Ursa Major	2	8.4	385
Polaris	Ursa Minor	2	2.0	430
Zeta (ζ) Boötis	Boötes	2	3.8	180

VARIABLE STARS TO OBSERVE

NAME	VARIABLE TYPE	CONSTELLATION	MINIMUM APPARENT MAGNITUDE	MAXIMUM APPARENT MAGNITUDE	PERIOD	DISTANCE FROM EARTH (LIGHT-YEARS)
Rasalgethi	Eclipsing binary	Hercules	4.1	3.1	128 days	382
Delta (δ) Cephei	Pulsating variable	Cepheus	4.4	3.5	5.3 days	982
Eta (η) Aquilae	Pulsating variable	Aquila	3.9	3.5	7.2 days	1,173
Eta (η) Geminorum	Eclipsing binary	Gemini	4.2	3.3	233 days	349
Gamma (γ) Cassiopeiae	Irregular variable	Cassiopeia	3.0	1.6	variable	613
Lambda (λ) Tauri	Eclipsing binary	Taurus	3.9	3.4	4 days	370
Mira	Pulsating variable	Cetus	10.0	2.0	332 days	418
Mu (μ) Cephei	Pulsating variable	Cepheus	5.1	3.4	730 or 4,400 days	5,258
W Virginis	Cepheid variable	Virgo	10.8	9.6	17 days	10,000
R Coronae Borealis	Irregular variable	Corona Borealis	14.8	5.8	variable	6,037
RR Lyrae	Pulsating variable	Lyra	8.1	7.1	0.6 days	744
Algol	Eclipsing binary	Perseus	3.4	2.1	2.9 days	93
Zeta (ζ) Geminorum	Pulsating variable	Gemini	4.2	3.6	10.2 days	1,168

THE CONSTELLATIONS (RANKED BY SIZE)

RANK	NAME	ABBREVIATION	AREA (SQ°)	NAMED BY
1	Hydra	Hya	1,303	Ptolemy
2	Virgo	Vir	1,294	Ptolemy
3	Ursa Major	UMa	1,280	Ptolemy
4	Cetus	Cet	1,231	Ptolemy
5	Hercules	Her	1,225	Ptolemy
6	Eridanus	Eri	1,138	Ptolemy
7	Pegasus	Peg	1,121	Ptolemy
8	Draco	Dra	1,082	Ptolemy
9	Centaurus	Cen	1,060	Ptolemy
10	Aquarius	Aqr	980	Ptolemy
11	Ophiuchus	Oph	948	Ptolemy
12	Leo	Leo	947	The Babylonians
13	Boötes	Boo	907	Ptolemy
14	Pisces	Psc	889	Ptolemy
15	Sagittarius	Sgr	867	Ptolemy
16	Cygnus	Cyg	803	Ptolemy
17	Taurus	Tau	797	Ptolemy
18	Camelopardalis	Cam	757	Petrus Plancius
19	Andromeda	And	722	Ptolemy
20	Puppis	Pup	673	Nicolas de Lacaille
21	Auriga	Aur	657	Ptolemy
22	Aquila	Aql	652	Ptolemy
23	Serpens	Ser	637	Ptolemy
24	Perseus	Per	615	Ptolemy
25	Cassiopeia	Cas	598	Ptolemy
26	Orion	Ori	594	Ptolemy
27	Cepheus	Cep	588	Ptolemy
28	Lynx	Lyn	545	Johannes Hevelius
29	Libra	Lib	538	The Romans
30	Gemini	Gem	514	Ptolemy
31	Cancer	Cnc	506	Ptolemy
32	Vela	Vel	500	Nicolas de Lacaille
33	Scorpius	Sco	497	Ptolemy
34	Carina	Car	494	Nicolas de Lacaille
35	Monoceros	Mon	482	Petrus Plancius
36	Sculptor	Scl	475	Nicolas de Lacaille
37	Phoenix	Phe	469	Keyser/De Houtman

THE CONSTELLATIONS CONTD

RANK	NAME	ABBREVIATION	AREA (SQ°)	NAMED BY
38	Canes Venatici	CVn	465	Johannes Hevelius
39	Aries	Ari	441	Ptolemy
40	Capricornus	Cap	414	The Babylonians
41	Fornax	For	398	Nicolas de Lacaille
42	Coma Berenices	Com	386	Caspar Vopel
43	Canis Major	CMa	380	Ptolemy
44	Pavo	Pav	378	Keyser/De Houtman
45	Grus	Gru	365	Keyser/De Houtman
46	Lupus	Lup	334	Ptolemy
47	Sextans	Sex	314	Johannes Hevelius
48	Tucana	Tuc	295	Keyser/De Houtman
49	Indus	Ind	294	Keyser/De Houtman
50	Octans	Oct	291	Nicolas de Lacaille
51	Lepus	Lep	290	Ptolemy
52	Lyra	Lyr	286	Ptolemy
53	Crater	Crt	282	Ptolemy
54	Columba	Col	270	Petrus Plancius
55	Vulpecula	Vul	268	Johannes Hevelius
56	Ursa Minor	UMi	255	Ptolemy
57	Telescopium	Tel	252	Nicolas de Lacaille
58	Horologium	Hor	252	Nicolas de Lacaille
59	Pictor	Pic	247	Nicolas de Lacaille
60	Piscis Austrinus	PsA	245	Ptolemy
61	Hydrus	Hyi	243	Ptolemy
62	Antlia	Ant	239	Nicolas de Lacaille
63	Ara	Ara	237	Ptolemy
64	Leo Minor	LMi	232	Johannes Hevelius
65	Pyxis	Pyx	221	Nicolas de Lacaille
66	Microscopium	Mic	210	Nicolas de Lacaille
67	Apus	Aps	206	Keyser/De Houtman
68	Lacerta	Lac	201	Johannes Hevelius
69	Delphinus	Del	189	Ptolemy
70	Corvus	Crv	184	Ptolemy
71	Canis Minor	CMi	183	Ptolemy
72	Dorado	Dor	179	Keyser/De Houtman
73	Corona Borealis	CrB	179	Ptolemy
74	Norma	Nor	165	Nicolas de Lacaille

THE CONSTELLATIONS CONTD

RANK	NAME	ABBREVIATION	AREA (SQ°)	NAMED BY
75	Mensa	Men	153	Nicolas de Lacaille
76	Volans	Vol	141	Keyser/De Houtman
77	Musca	Mus	138	Keyser/De Houtman
78	Triangulum	Tri	132	Ptolemy
79	Chamaeleon	Cha	132	Keyser/De Houtman
80	Corona Australis	Cra	128	Ptolemy
81	Caelum	Cae	125	Nicolas de Lacaille
82	Reticulum	Ret	114	Nicolas de Lacaille
83	Triangulum Australe	TrA	110	Keyser/De Houtman
84	Scutum	Sct	109	Johannes Hevelius
85	Cir	Circinus	93	Nicolas de Lacaille
86	Sagitta	Sge	80	Ptolemy
87	Equuleus	Equ	72	Ptolemy
88	Crux	Cru	68	Augustin Royer

THE CONSTELLATION CRUX

GALAXIES TO OBSERVE

NAME/ CATALOG NUMBER	GALAXY TYPE	CONSTELLATION	APPARENT MAGNITUDE	DISTANCE FROM EARTH (LIGHT-YEARS)
Andromeda Galaxy (M31)	Spiral	Andromeda	4.5	2.5 million
Black Eye Galaxy (M64)	Spiral	Coma Berenices	8.5	17 million
Bode's Galaxy (M81)	Spiral	Ursa Major	6.9	12 million
Cigar Galaxy (M82)	Spiral	Ursa Major	8.4	12 million
Large Magellanic Cloud	Irregular	Dorado	0.4	179,000
Small Magellanic Cloud (NGC 292)	Irregular	Tucana	2.3	200,000
Sombrero Galaxy (M104)	Spiral	Virgo	9.0	30 million
Triangulum Galaxy (M33)	Spiral	Triangulum	5.7	3 million
Whirlpool Galaxy (M51)	Spiral	Canes Venatici	8.4	23 million

NEBULAE TO OBSERVE

NAME	NEBULA TYPE	CONSTELLATION	APPARENT MAGNITUDE	DISTANCE FROM EARTH (LY)
Cat's Eye Nebula	Planetary	Draco	8.1	3,600
Cone Nebula	Dark	Monoceros	3.9	2,500
Crescent Nebula	Planetary	Cygnus	7.4	4,700
Dumbell Nebula	Planetary	Vulpecula	7.6	1,000
Eagle Nebula	Emission	Serpens Cauda	6.0	7,000
Eight-Burst Nebula	Planetary	Vela	8.1	2,000
Eskimo Nebula	Planetary	Gemini	8.6	3,800
Eta Carinae Nebula	Emission	Carina	1.0	7,500
Helix Nebula	Planetary	Aquarius	6.5	650
Hourglass Nebula	Planetary	Musca	11.8	8,000
IC 2944	Emission	Centaurus	4.5	5,900
Lagoon Nebula	Emission	Sagittarius	5.8	5,200
Orion Nebula	Emission	Orion	4.0	1,350
Owl Nebula	Planetary	Ursa Major	9.8	2,600
Ring Nebula	Planetary	Lyra	9.0	2,000
Saturn Nebula	Planetary	Aquarius	8.0	5,200
Stingray Nebula	Planetary	Ara	10.8	18,000
Tarantula Nebula	Emission	Dorado	5.0	160,000
Trifid Nebula	Emission	Sagittarius	9.0	7,000
Veil Nebula	Supernova remnant	Cygnus	7.0	2,000

MESSIER CATALOG

MESSIER NUMBER	CONSTELLATION	COMMON NAME	OBJECT TYPE
M1	Taurus	Crab Nebula	Supernova remnant
M2	Aquarius	–	Globular cluster
M3	Canes Venatici	–	Globular cluster
M4	Scorpius	–	Globular cluster
M5	Serpens Caput	–	Globular cluster
M6	Scorpius	Butterfly Cluster	Open cluster
M7	Scorpius	Ptolemy Cluster	Open cluster
M8	Sagittarius	Lagoon Nebula	Emission nebula
M9	Ophiuchus	–	Globular cluster
M10	Ophiuchus	–	Globular cluster
M11	Scutum	Wild Duck Cluster	Open cluster
M12	Ophiuchus	–	Globular cluster
M13	Hercules	The Great Globular	Globular cluster
M14	Ophiuchus	–	Globular cluster
M15	Pegasus	–	Globular cluster
M16	Serpens Cauda	The Eagle Nebula	Open cluster/ emission nebula
M17	Sagittarius	Omega/Swan Nebula	Emission nebula
M18	Sagittarius	–	Open cluster
M19	Ophiuchus	–	Globular cluster
M20	Sagittarius	Trifid Nebula	Emission/reflection/ dark nebula
M21	Sagittarius	–	Open cluster
M22	Sagittarius	–	Globular cluster
M23	Sagittarius	–	Open cluster
M24	Sagittarius	Sagittarius Star Cloud	Starfield
M25	Sagittarius	–	Open cluster
M26	Scutum	–	Open cluster
M27	Vulpecula	Dumbbell Nebula	Planetary nebula
M28	Sagittarius	–	Globular cluster
M29	Cygnus	–	Open cluster
M30	Capricornus	–	Globular cluster

MESSIER CATALOG CONTD

MESSIER NUMBER	CONSTELLATION	COMMON NAME	OBJECT TYPE
M31	Andromeda	Andromeda Galaxy	Spiral galaxy
M32	Andromeda	–	Dwarf elliptical galaxy
M33	Triangulum	Triangulum Galaxy	Spiral galaxy
M34	Perseus	–	Open cluster
M35	Gemini	–	Open cluster
M36	Auriga	–	Open cluster
M37	Auriga	–	Open cluster
M38	Auriga	–	Open cluster
M39	Cygnus	–	Open cluster
M40	Ursa Major	Winnecke 4	Double star
M41	Canis Major	–	Open cluster
M42	Orion	Orion Nebula	Emission/reflection nebula
M43	Orion	De Mairan's Nebula	Emission/reflection nebula
M44	Cancer	Beehive Cluster	Open cluster
M45	Taurus	Pleiades/Seven Sisters	Open cluster
M46	Puppis	–	Open cluster
M47	Puppis	–	Open cluster
M48	Hydra	–	Open cluster
M49	Virgo	–	Elliptical galaxy
M50	Monoceros	–	Open cluster
M51	Canes Venatici	Whirlpool Galaxy	Spiral galaxy
M52	Cassiopeia	–	Open cluster
M53	Coma Berenices	–	Globular cluster
M54	Sagittarius	–	Globular cluster
M55	Sagittarius	–	Globular cluster
M56	Lyra	–	Globular cluster
M57	Lyra	Ring Nebula	Planetary nebula
M58	Virgo	–	Barred spiral galaxy
M59	Virgo	–	Elliptical galaxy
M60	Virgo	–	Elliptical galaxy
M61	Virgo	–	Spiral galaxy
M62	Ophiuchus	–	Globular cluster

MESSIER CATALOG CONTD

MESSIER NUMBER	CONSTELLATION	COMMON NAME	OBJECT TYPE
M63	Canes Venatici	Sunflower Galaxy	Spiral galaxy
M64	Coma Berenices	Black Eye Galaxy	Spiral galaxy
M65	Leo	–	Spiral galaxy
M66	Leo	–	Spiral galaxy
M67	Cancer	–	Open cluster
M68	Hydra	–	Globular cluster
M69	Sagittarius	–	Globular cluster
M70	Sagittarius	–	Globular cluster
M71	Sagitta	–	Globular cluster
M72	Aquarius	–	Globular cluster
M73	Aquarius	–	Asterism
M74	Pisces	–	Spiral galaxy
M75	Sagittarius	–	Globular cluster
M76	Perseus	Little Dumbbell Nebula	Planetary nebula
M77	Cetus	–	Barred spiral galaxy
M78	Orion	–	Reflection nebula
M79	Lepus	–	Globular cluster
M80	Scorpius	–	Globular cluster
M81	Ursa Major	Bode's Galaxy	Spiral galaxy
M82	Ursa Major	Cigar Galaxy	Spiral galaxy
M83	Hydra	Southern Pinwheel Galaxy	Barred spiral galaxy
M84	Virgo	–	Lenticular galaxy
M85	Coma Berenices	–	Lenticular galaxy
M86	Virgo	–	Lenticular galaxy
M87	Virgo	Virgo A	Elliptical galaxy
M88	Coma Berenices	–	Spiral galaxy
M89	Virgo	–	Elliptical galaxy
M90	Virgo	–	Spiral galaxy
M91	Coma Berenices	–	Barred spiral galaxy
M92	Hercules	–	Globular cluster

MESSIER CATALOG CONTD

MESSIER NUMBER	CONSTELLATION	COMMON NAME	OBJECT TYPE
M93	Puppis	–	Open cluster
M94	Canes Venatici	–	Spiral galaxy
M95	Leo	–	Barred spiral galaxy
M96	Leo	–	Spiral galaxy
M97	Ursa Major	Owl Nebula	Planetary nebula
M98	Coma Berenices	–	Spiral galaxy
M99	Coma Berenices	–	Spiral galaxy
M100	Coma Berenices	–	Spiral galaxy
M101	Ursa Major	Pinwheel Galaxy	Spiral galaxy
M102	Not unambiguously identified	–	–
M103	Cassiopeia	–	Open cluster
M104	Virgo	Sombrero Galaxy	Unbarred spiral galaxy
M105	Leo	–	Elliptical galaxy
M106	Canes Venatici	–	Spiral galaxy
M107	Ophiuchus	–	Globular cluster
M108	Ursa Major	–	Barred spiral galaxy
M109	Ursa Major	–	Barred spiral galaxy
M110	Andromeda	–	Dwarf elliptical galaxy

M8, THE LAGOON NEBULA

ALMANAC

There is always something happening in the sky, but there are times when it is worth taking a closer look—especially if it is for a rare event or something you may never have seen before. We look here at the next few years, including such events as solar and lunar eclipses, and the best time to look at the planets.

KEY	
⬤	Full Moon
⬤	New Moon
⬤	Total eclipse of the Moon
⬤	Partial eclipse of the Moon
⬤	Total eclipse of the Sun
⬤	Partial eclipse of the Sun
⬤	Annular eclipse

Opposition and conjunction

When one of the inferior planets, Mercury or Venus, lies directly between us and the Sun, this is known as inferior conjunction. When either is directly behind the Sun from our point of view, this position is called superior conjunction. A planet at conjunction is lost in the Sun's glare (see p.176). Inferior planets are said to be at greatest elongation when they are at their farthest visible point from the Sun and thus are at their best viewing points. Looking directly away from the Sun, when a superior planet lies directly opposite the Sun in the sky, it is said to be at opposition. This means the planet is visible most of the night and is at its brightest and highest in the sky.

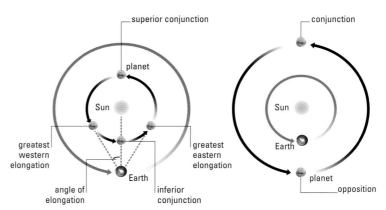

INFERIOR ORBIT
Mercury and Venus together are known as the inferior planets, which simply means they orbit closer to the Sun than Earth.

SUPERIOR ORBIT
The major planets that orbit farther from the Sun than Earth are called the superior planets. They are Mars, Jupiter, Saturn, Uranus, and Neptune.

Eclipses

When the orbits of the Sun and Moon align, relative to Earth, we are treated to an eclipse. With its tilted orbit around us, the Moon only occasionally passes directly between us and the Sun, and at this time we witness a solar eclipse (see p.193). If the whole Moon covers the Sun, it is known as a total eclipse. If the Moon passes across the center of the Sun but is too far from Earth to cover the Sun completely, it is called an annular eclipse. A partial eclipse occurs when the Moon only covers part of the Sun. The Moon may line up directly opposite the Sun and move into Earth's shadow, producing a lunar eclipse (see p.197).

SOLAR ECLIPSE MAP
Here are the dates and paths of eclipses around the world. Note there is one hybrid eclipse on April 20, 2023, which starts annular, becomes total, and returns to annular.

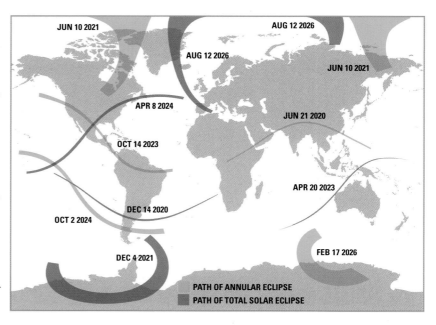

2020

Chile and Argentina are extremely fortunate: last year, in July, they saw a total eclipse of the Sun, and now they have a second one in December of this year. In the night sky, Mars puts on a fine opposition show in October, where it will be shining brightly in the constellation Pisces. Earlier in the year, Venus will be well placed for evening observation in March.

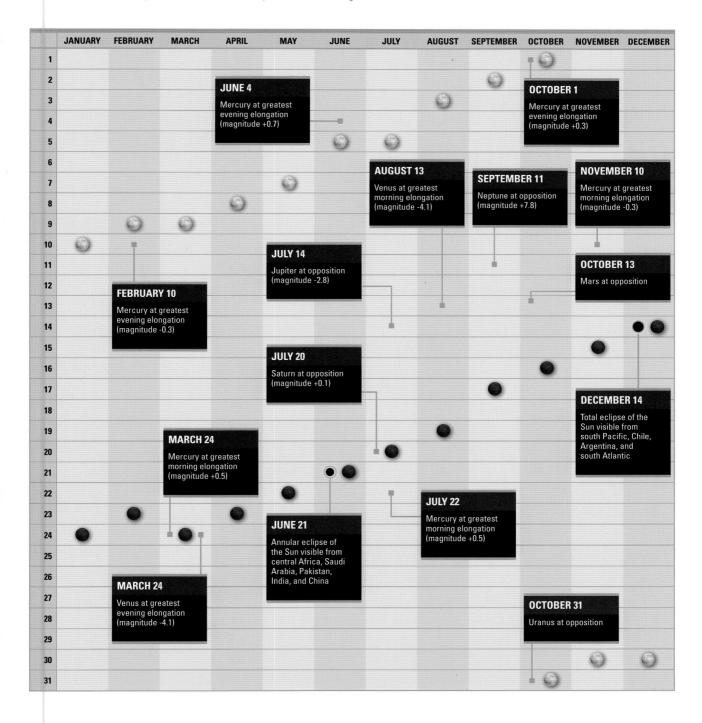

JUNE 4
Mercury at greatest evening elongation (magnitude +0.7)

OCTOBER 1
Mercury at greatest evening elongation (magnitude +0.3)

AUGUST 13
Venus at greatest morning elongation (magnitude -4.1)

SEPTEMBER 11
Neptune at opposition (magnitude +7.8)

NOVEMBER 10
Mercury at greatest morning elongation (magnitude -0.3)

JULY 14
Jupiter at opposition (magnitude -2.8)

OCTOBER 13
Mars at opposition

FEBRUARY 10
Mercury at greatest evening elongation (magnitude -0.3)

JULY 20
Saturn at opposition (magnitude +0.1)

DECEMBER 14
Total eclipse of the Sun visible from south Pacific, Chile, Argentina, and south Atlantic

MARCH 24
Mercury at greatest morning elongation (magnitude +0.5)

JUNE 21
Annular eclipse of the Sun visible from central Africa, Saudi Arabia, Pakistan, India, and China

JULY 22
Mercury at greatest morning elongation (magnitude +0.5)

MARCH 24
Venus at greatest evening elongation (magnitude -4.1)

OCTOBER 31
Uranus at opposition

2021

There are four eclipses this year: an annular and a total solar eclipse, around the higher northern latitudes and higher southern latitudes, respectively; and a partial and total lunar eclipse, both centered around the Pacific. As for the planets, Venus looks good in the evening skies of October, and Jupiter will be shining brightly during its opposition in Aquarius at the end of August.

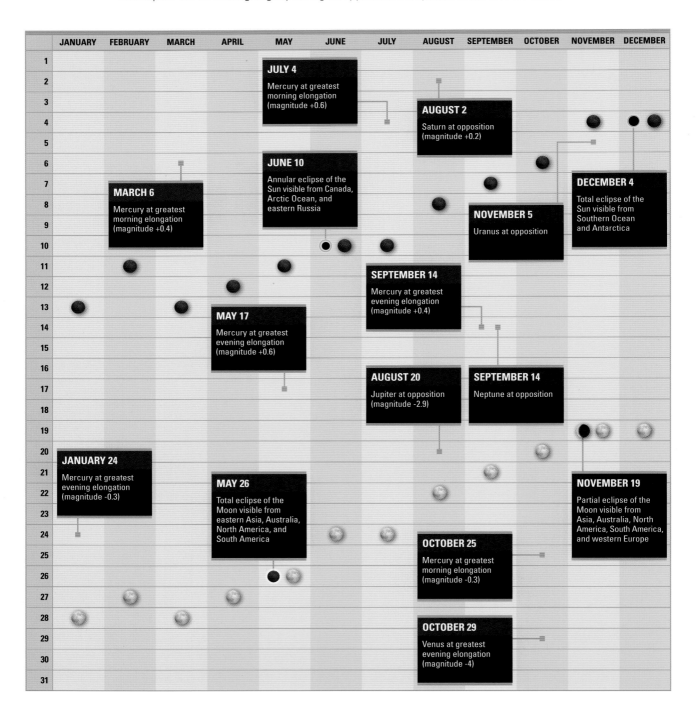

	JANUARY	FEBRUARY	MARCH	APRIL	MAY	JUNE	JULY	AUGUST	SEPTEMBER	OCTOBER	NOVEMBER	DECEMBER

JULY 4
Mercury at greatest morning elongation (magnitude +0.6)

AUGUST 2
Saturn at opposition (magnitude +0.2)

MARCH 6
Mercury at greatest morning elongation (magnitude +0.4)

JUNE 10
Annular eclipse of the Sun visible from Canada, Arctic Ocean, and eastern Russia

DECEMBER 4
Total eclipse of the Sun visible from Southern Ocean and Antarctica

NOVEMBER 5
Uranus at opposition

SEPTEMBER 14
Mercury at greatest evening elongation (magnitude +0.4)

MAY 17
Mercury at greatest evening elongation (magnitude +0.6)

AUGUST 20
Jupiter at opposition (magnitude -2.9)

SEPTEMBER 14
Neptune at opposition

JANUARY 24
Mercury at greatest evening elongation (magnitude -0.3)

MAY 26
Total eclipse of the Moon visible from eastern Asia, Australia, North America, and South America

NOVEMBER 19
Partial eclipse of the Moon visible from Asia, Australia, North America, South America, and western Europe

OCTOBER 25
Mercury at greatest morning elongation (magnitude -0.3)

OCTOBER 29
Venus at greatest evening elongation (magnitude -4)

2022

It isn't until August that we can start to see any planets in the evening sky. Until then, there is only really Venus to look at, shining brightly in the morning sky and reaching greatest eastern elongation on March 20. However, if you look closely around this time, you will see two "stars" close to Venus, which are in fact the planets Mars and Saturn.

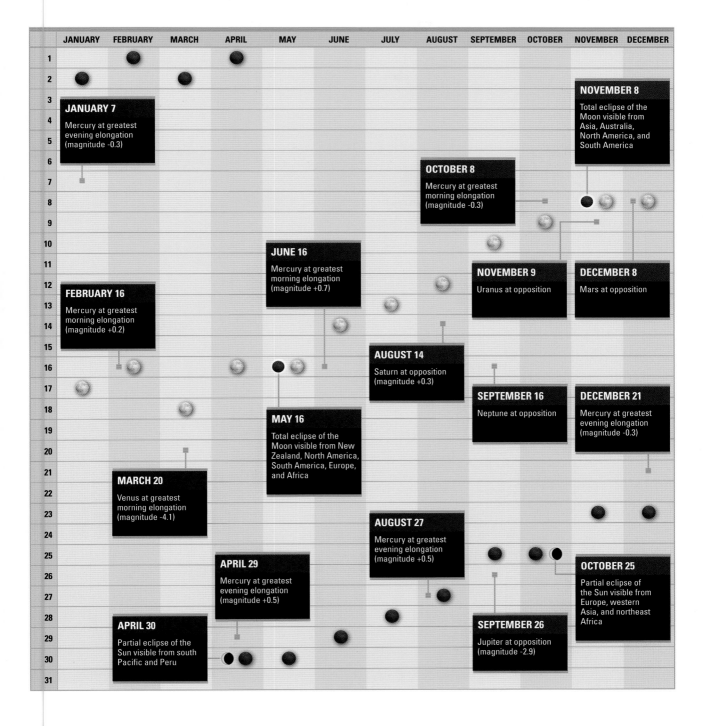

JANUARY 7
Mercury at greatest evening elongation (magnitude -0.3)

FEBRUARY 16
Mercury at greatest morning elongation (magnitude +0.2)

MARCH 20
Venus at greatest morning elongation (magnitude -4.1)

APRIL 29
Mercury at greatest evening elongation (magnitude +0.5)

APRIL 30
Partial eclipse of the Sun visible from south Pacific and Peru

MAY 16
Total eclipse of the Moon visible from New Zealand, North America, South America, Europe, and Africa

JUNE 16
Mercury at greatest morning elongation (magnitude +0.7)

AUGUST 14
Saturn at opposition (magnitude +0.3)

AUGUST 27
Mercury at greatest evening elongation (magnitude +0.5)

SEPTEMBER 16
Neptune at opposition

SEPTEMBER 26
Jupiter at opposition (magnitude -2.9)

OCTOBER 8
Mercury at greatest morning elongation (magnitude -0.3)

OCTOBER 25
Partial eclipse of the Sun visible from Europe, western Asia, and northeast Africa

NOVEMBER 8
Total eclipse of the Moon visible from Asia, Australia, North America, and South America

NOVEMBER 9
Uranus at opposition

DECEMBER 8
Mars at opposition

DECEMBER 21
Mercury at greatest evening elongation (magnitude -0.3)

2023

The solar eclipse of April 20 is a rare occurrence, a mixture of an annular and a total eclipse known as a hybrid solar eclipse. Over the course of this type of eclipse, the Moon only comes close enough to Earth to produce totality around the middle of the eclipse. At the start and finish, it is farther away and produces an annular eclipse.

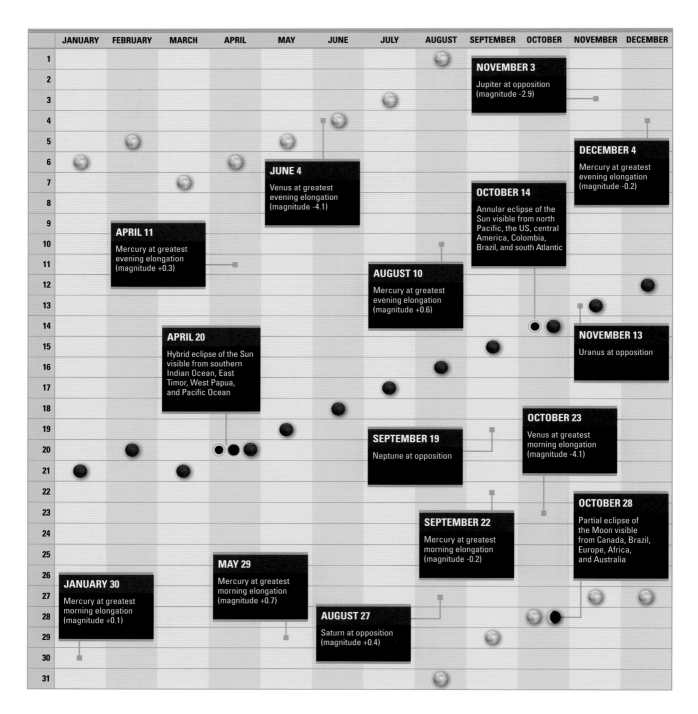

	JANUARY	FEBRUARY	MARCH	APRIL	MAY	JUNE	JULY	AUGUST	SEPTEMBER	OCTOBER	NOVEMBER	DECEMBER

NOVEMBER 3
Jupiter at opposition (magnitude -2.9)

JUNE 4
Venus at greatest evening elongation (magnitude -4.1)

DECEMBER 4
Mercury at greatest evening elongation (magnitude -0.2)

OCTOBER 14
Annular eclipse of the Sun visible from north Pacific, the US, central America, Colombia, Brazil, and south Atlantic

APRIL 11
Mercury at greatest evening elongation (magnitude +0.3)

AUGUST 10
Mercury at greatest evening elongation (magnitude +0.6)

NOVEMBER 13
Uranus at opposition

APRIL 20
Hybrid eclipse of the Sun visible from southern Indian Ocean, East Timor, West Papua, and Pacific Ocean

SEPTEMBER 19
Neptune at opposition

OCTOBER 23
Venus at greatest morning elongation (magnitude -4.1)

OCTOBER 28
Partial eclipse of the Moon visible from Canada, Brazil, Europe, Africa, and Australia

SEPTEMBER 22
Mercury at greatest morning elongation (magnitude -0.2)

MAY 29
Mercury at greatest morning elongation (magnitude +0.7)

JANUARY 30
Mercury at greatest morning elongation (magnitude +0.1)

AUGUST 27
Saturn at opposition (magnitude +0.4)

2024

One of the main events of this year is a total solar eclipse moving across North America on April 8, mirroring a similar event that occurred in 2017. Sometimes—due to the alignment of the Sun, Earth, and Moon—certain places are lucky enough to see this spectacular event within a few years of one another. The path of this year's event will make it one of the most viewed solar eclipses of the decade.

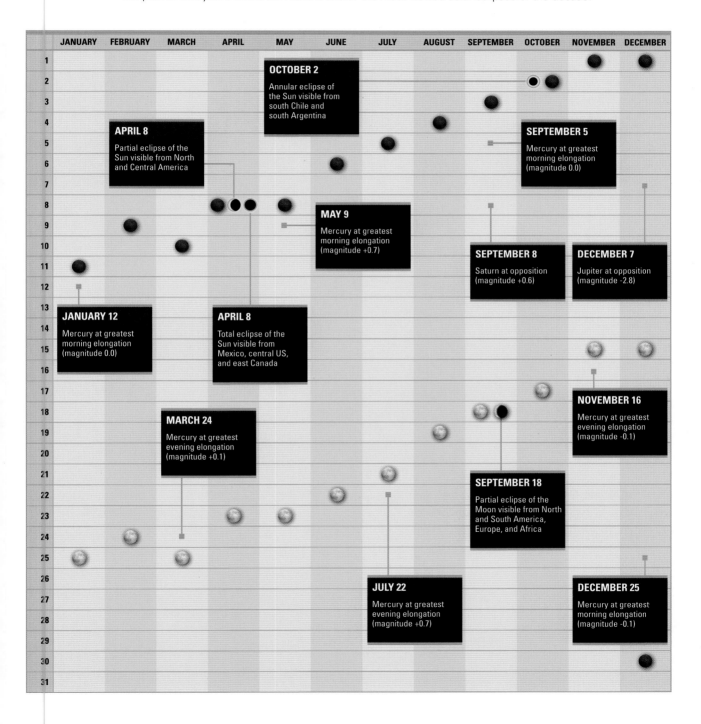

	JANUARY	FEBRUARY	MARCH	APRIL	MAY	JUNE	JULY	AUGUST	SEPTEMBER	OCTOBER	NOVEMBER	DECEMBER

OCTOBER 2
Annular eclipse of the Sun visible from south Chile and south Argentina

APRIL 8
Partial eclipse of the Sun visible from North and Central America

SEPTEMBER 5
Mercury at greatest morning elongation (magnitude 0.0)

MAY 9
Mercury at greatest morning elongation (magnitude +0.7)

SEPTEMBER 8
Saturn at opposition (magnitude +0.6)

DECEMBER 7
Jupiter at opposition (magnitude -2.8)

JANUARY 12
Mercury at greatest morning elongation (magnitude 0.0)

APRIL 8
Total eclipse of the Sun visible from Mexico, central US, and east Canada

MARCH 24
Mercury at greatest evening elongation (magnitude +0.1)

NOVEMBER 16
Mercury at greatest evening elongation (magnitude -0.1)

SEPTEMBER 18
Partial eclipse of the Moon visible from North and South America, Europe, and Africa

JULY 22
Mercury at greatest evening elongation (magnitude +0.7)

DECEMBER 25
Mercury at greatest morning elongation (magnitude -0.1)

2025

In the late-evening skies at the start of the year, it is worth looking for two planets: Jupiter is unmistakably the brightest object in the sky, and off to its east is the fainter but still bright planet Mars. This year also sees two total lunar eclipses—the one on March 14 is visible in North and South America, while Africa, central and eastern Europe, Asia, and Australia see another lunar eclipse on September 7.

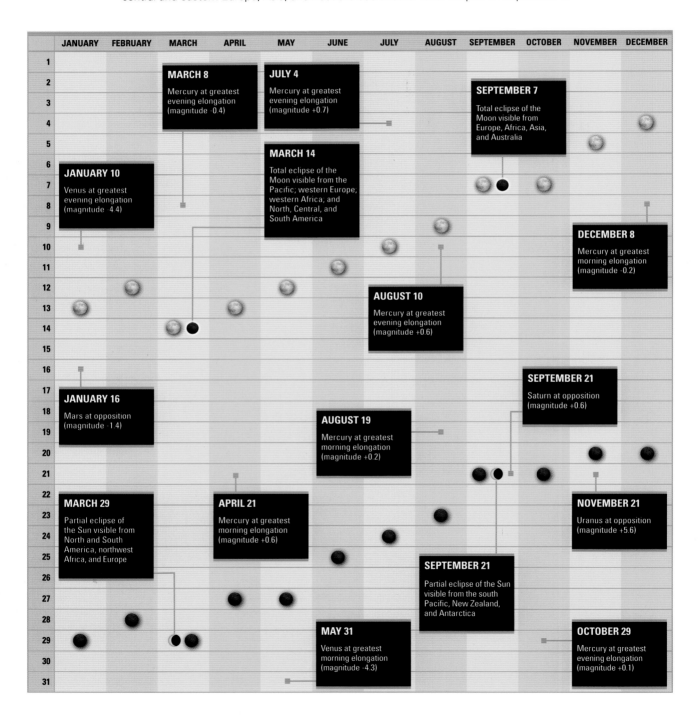

	JANUARY	FEBRUARY	MARCH	APRIL	MAY	JUNE	JULY	AUGUST	SEPTEMBER	OCTOBER	NOVEMBER	DECEMBER

MARCH 8
Mercury at greatest evening elongation (magnitude -0.4)

JULY 4
Mercury at greatest evening elongation (magnitude +0.7)

SEPTEMBER 7
Total eclipse of the Moon visible from Europe, Africa, Asia, and Australia

MARCH 14
Total eclipse of the Moon visible from the Pacific; western Europe; western Africa; and North, Central, and South America

JANUARY 10
Venus at greatest evening elongation (magnitude -4.4)

DECEMBER 8
Mercury at greatest morning elongation (magnitude -0.2)

AUGUST 10
Mercury at greatest evening elongation (magnitude +0.6)

JANUARY 16
Mars at opposition (magnitude -1.4)

SEPTEMBER 21
Saturn at opposition (magnitude +0.6)

AUGUST 19
Mercury at greatest morning elongation (magnitude +0.2)

MARCH 29
Partial eclipse of the Sun visible from North and South America, northwest Africa, and Europe

APRIL 21
Mercury at greatest morning elongation (magnitude +0.6)

NOVEMBER 21
Uranus at opposition (magnitude +5.6)

SEPTEMBER 21
Partial eclipse of the Sun visible from the south Pacific, New Zealand, and Antarctica

MAY 31
Venus at greatest morning elongation (magnitude -4.3)

OCTOBER 29
Mercury at greatest evening elongation (magnitude +0.1)

2026

As in the previous year, Jupiter starts the year as the brightest "star" in the evening sky. In mid-August, in the early evening of August 12, a total solar eclipse crosses northwest Spain to reach the Balearics just before sunset. This means that the night sky is Moon-free for the annual meteor shower, the Perseids, that occurs later that same evening (and on the following day, too).

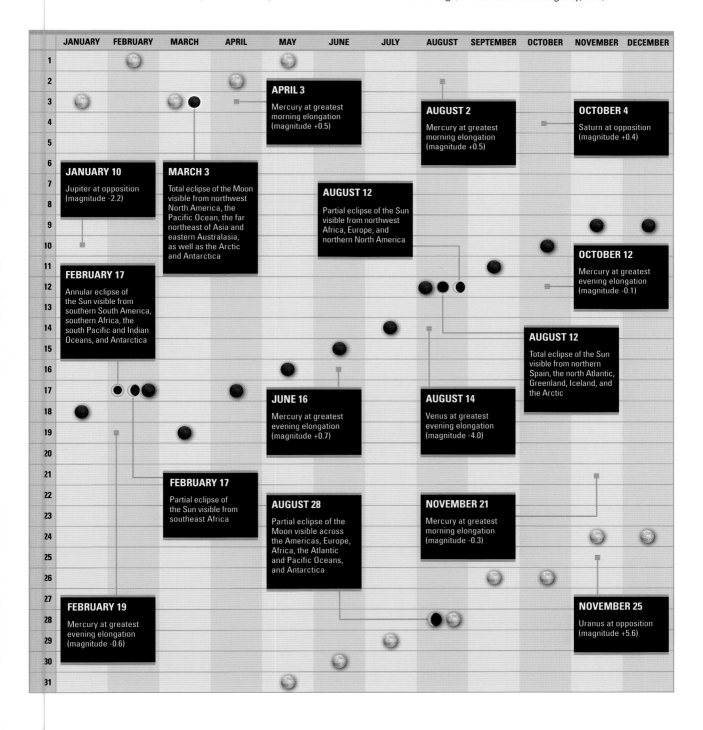

| | JANUARY | FEBRUARY | MARCH | APRIL | MAY | JUNE | JULY | AUGUST | SEPTEMBER | OCTOBER | NOVEMBER | DECEMBER |

APRIL 3
Mercury at greatest morning elongation (magnitude +0.5)

AUGUST 2
Mercury at greatest morning elongation (magnitude +0.5)

OCTOBER 4
Saturn at opposition (magnitude +0.4)

JANUARY 10
Jupiter at opposition (magnitude -2.2)

MARCH 3
Total eclipse of the Moon visible from northwest North America, the Pacific Ocean, the far northeast of Asia and eastern Australasia, as well as the Arctic and Antarctica

AUGUST 12
Partial eclipse of the Sun visible from northwest Africa, Europe, and northern North America

OCTOBER 12
Mercury at greatest evening elongation (magnitude -0.1)

FEBRUARY 17
Annular eclipse of the Sun visible from southern South America, southern Africa, the south Pacific and Indian Oceans, and Antarctica

AUGUST 12
Total eclipse of the Sun visible from northern Spain, the north Atlantic, Greenland, Iceland, and the Arctic

JUNE 16
Mercury at greatest evening elongation (magnitude +0.7)

AUGUST 14
Venus at greatest evening elongation (magnitude -4.0)

FEBRUARY 17
Partial eclipse of the Sun visible from southeast Africa

AUGUST 28
Partial eclipse of the Moon visible across the Americas, Europe, Africa, the Atlantic and Pacific Oceans, and Antarctica

NOVEMBER 21
Mercury at greatest morning elongation (magnitude -0.3)

FEBRUARY 19
Mercury at greatest evening elongation (magnitude -0.6)

NOVEMBER 25
Uranus at opposition (magnitude +5.6)

GLOSSARY

Albedo A measure of how reflective the surface of an object is. Astronomers sometimes refer to albedo features on planetary surfaces; these dark or light markings are visible through telescopes.

Alt-azimuth mount A type of telescope mounting that allows a telescope to be moved around 360° horizontally (in azimuth) and between level and 90° vertically (in altitude).

Altitude A measure of how far (in degrees) an object is above the observer's horizon.

Apparent magnitude The brightness of a celestial object as perceived by an observer standing on Earth.

Aperture The size of a telescope's main mirror or lens (often called the primary mirror or objective lens). Larger-aperture telescopes can collect more light and are better for observing faint objects.

Asteroid A small irregular Solar System object, typically ranging in size from a few feet to several hundreds of miles across. Asteroids are made of rock and/or metal and are thought to be the detritus left over from the formation of the planets.

Aurora A phenomenon of the Earth's upper atmosphere (and the atmospheres of some other planets). When a burst of energetic particles from the Sun disturbs Earth's magnetic field, electrically charged particles trapped by it are funneled down close to the polar regions. They collide with gas atoms in the atmosphere, creating glowing displays of colored light.

Azimuth A measure of distance around the horizon, measured in degrees, from the zero point of due north.

Barlow lens A telescope accessory that is used in conjunction with an eyepiece to increase the magnification, typically by two or three times.

Binary star A pair of stars, gravitationally linked, which orbit a common center of mass (known as a barycenter). Many binary stars can be distinguished with a small telescope.

Cable release An accessory used with a camera to allow remote shutter firing and prevent camera vibration.

CCD camera Charge Coupled Device camera; a specialized type of camera used for astro-imaging. The CCD uses a sensitive sensor chip to convert photons of light falling on it into an electrical signal. This signal is then read by the camera's electronics to produce the image.

Celestial sphere The imaginary sphere onto which the stars and constellations are mapped. Coordinates on the celestial sphere are similar to latitude and longitude on Earth, with the respective celestial coordinates designating declination (DEC) and right ascension (RA).

Circumpolar A star that, when seen from one particular location, does not sink below the horizon over the course of the night.

Cloud band A belt of cloud surrounding a giant planet parallel to its equator.

Comet An icy celestial body, originating as a solid nucleus in a vast reservoir in the outer Solar System. When a comet nucleus passes into the inner Solar System, solar heating causes it to develop a coma and tails of gas and dust.

Constellation A named pattern of stars or a designated area of sky around a star pattern. There are currently 88 officially recognized constellations.

Corona The outer atmosphere of the Sun. It can be seen during a total solar eclipse as a halo of wispy streamers emanating from the Sun's disc. The corona is much hotter than the surface of the Sun, with a temperature of around 3.6 million °F.

Cosmic Microwave Background The radiation left over from the Big Bang, appearing from all directions in the sky. Today it appears as microwave radiation, but when it was formed, it was much higher-energy short-wave radiation.

Declination (DEC) A coordinate used in the equatorial coordinate system. It is measured in degrees above or below the celestial equator, which has a declination of 0°.

Dew shield A cylindrical tube extension placed on the front of telescopes to prevent dew from condensing onto the corrector plate or lens and degrading the view through the eyepiece.

DSLR Digital Single Lens Reflex; a type of camera often used for astro-imaging because of its versatility, which makes it useful for capturing the light from faint galaxies or nebulae.

Ecliptic The path on the celestial sphere that the Sun traces across the sky during a year.

Electromagnetic spectrum The entire range of energy emitted by the different objects in the Universe. Our eyes can see a specific range within this spectrum, which we call "visible light."

Elliptical galaxy A galaxy that is elliptical in shape, appearing as a "blob" through a telescope and in images. Elliptical galaxies lack large amounts of gas and dust and therefore, unlike spiral galaxies, are also devoid of star formation.

Emission nebula A huge cloud of interstellar dust and gas that usually glows because of the radiation from a nearby star or group of stars. The Orion Nebula (M42) is an emission nebula.

Equatorial mount A type of telescope mount that allows the observer to align the mount accurately to the celestial sphere and so track the stars. An equatorial mount has a right ascension (RA) axis and a declination (DEC) axis and must be polar-aligned to work properly.

Eyepiece A telescope accessory that is used to magnify the image produced by a telescope. The number printed on the side of an eyepiece is the eyepiece's focal length.

Finderscope/finder A small telescope or device that is attached to the top of the main telescope. It is used to locate and center an object in the main telescope's field of view. Red dot finders do not actually magnify the view, but project a target such as a dot or set of circles to help center the object.

Focal length The distance between a telescope's primary mirror or objective lens and the point where the image is focused.

Focal ratio The number calculated when a telescope's focal length (in millimeters) is divided by its aperture (also in millimeters). Large focal-ratio telescopes are good for looking at bright objects, while small focal-ratio telescopes are more suited to observing fainter objects.

Focuser The assembly on a telescope that holds the eyepiece in place and is used to bring the image seen through the eyepiece into focus.

Galaxy A huge mass of stars linked by gravity. Types include elliptical, lenticular, irregular, and spiral galaxies such as our own Milky Way. Galaxies can range from a few thousand to hundreds of thousands of light-years across.

Globular cluster A sphere of stars, with tens or even hundreds of thousands of members bound together by gravity. Globular clusters are ancient objects that are many billions of years old.

Magnification The number of times the naked-eye view of an object is enlarged. The magnification of a given telescope and eyepiece combination is calculated by dividing the focal length of the telescope by that of the eyepiece.

Magnitude See *Apparent magnitude*.

Maksutov–Cassegrain A design of catadioptric telescope that uses a primary mirror, a secondary mirror, and a meniscus lens to collect light and bring it into focus. Maksutov–Cassegrains or "Maks" are generally quite compact and make good lunar or planetary telescopes.

Meteor The correct name for the streak of light that is sometimes called a "shooting star." A meteor is the result of a small particle entering Earth's atmosphere at very high speed. As it hits the atmosphere, it violently compresses the air ahead of it. The air quickly heats up and in turn heats the particle, too, causing it to glow.

Meteorite A meteoroid that reaches the ground and survives impact. They are usually classified according to their composition as stony, iron, or stony-iron.

Meteoroid A small piece of space rock up to boulder size. Meteoroids create meteors when they enter Earth's atmosphere.

Milky Way The name given to the spiral galaxy that is home to the Solar System. It is thought that the Milky Way is 180,000 light-years in diameter and contains somewhere between 200 and 400 billion stars.

Moon A natural satellite orbiting a planet. Most of the planets in the Solar System have a moon or family of moons orbiting them.

North and south celestial pole The two imaginary points in the sky where Earth's rotation axis intersects the celestial sphere itself. In the northern hemisphere, the star Polaris is very close to the north celestial pole.

Open cluster A loose group of stars that formed at the same time, typically consisting of several hundred or several thousand members. Open clusters are found in the arms of spiral galaxies, some within the nebulae where they were born.

Optical tube assembly The tube of a telescope that houses its lenses or mirrors. The optical tube assembly is often abbreviated to "OTA."

Orbit The path a celestial body takes in space under the influence of the gravity of other objects relatively nearby. Basic closed orbits, where one body goes around another, such as planets going around the Sun, are elliptical.

Photon A tiny packet of electromagnetic radiation, such as light. Though electromagnetic radiation travels like a wave, it also behaves as a stream of particles—photons. Astronomers aim to capture as many photons as they can from faint objects.

Photosphere The visible "surface" of the Sun or any other star. The hot gas is opaque beneath this layer but transparent above it. The temperature of the Sun's photosphere is about 9,930°F.

Planet A celestial body that has cleared away any planetary debris from its orbit around the Sun and is roughly round thanks to the effects of its own gravity. There are currently eight objects considered to be planets in the Solar System.

Planetary nebula A nebula formed when a star of a mass similar to the Sun begins to die. As it does so, it gently expels its outer layers, creating a glowing shell of dust and gas that is the planetary nebula. The term was invented by William Herschel because in his telescopes, they looked similar to planets.

Planisphere A star chart used for working out which constellations are visible at a given time and date from a set range of observing latitudes. It consists of two discs that can be rotated to display the stars visible (at a chosen time and date) through a clear window.

Plasma A mixture of electrons and ions that behaves like a gas but conducts electricity and is affected by magnetic fields. Plasma makes up at least 99 percent of the Universe.

Polaris Also called the "North Star" or the "Pole Star"; this is a circumpolar northern-hemisphere star within the constellation of Ursa Minor that is very close to the north celestial pole.

Prominence A vast ejection of plasma from the Sun's atmosphere. Prominences appear through special filtered telescopes as great red tendrils of gas reaching out from the Sun's photosphere.

Protoplanetary disc A disc of dusty, rocky, and gaseous material, usually orbiting a young star. It is within a protoplanetary disc that planets begin to form. Many protoplanetary discs have been observed around stars.

Pulsar A highly magnetized neutron star with a powerful magnetic field, formed when a massive star dies in a supernova. If the poles of the magnetic field are not aligned with the rotation axis of the neutron star, jets of radiation sweep around space at high speed, creating a pulsar.

Red giant When a star of similar size to our Sun begins to die, it swells hugely in size and cools, slowly becoming what is known as a red giant.

Reflection nebula A nebula containing tiny dust particles that reflect light toward Earth.

Right ascension (RA) A coordinate on the celestial sphere, used in the equatorial system. Right ascension is measured in hours from the position where the ecliptic intersects the celestial equator in spring.

Schmidt–Cassegrain A telescope that uses a primary mirror, secondary mirror, and corrector plate to gather light and produce an image.

Seeing A measure of how steady the atmosphere is. Poor seeing can make planetary and lunar observing particularly difficult, as the undulations in the Earth's atmosphere blur and distort fine detail seen through the eyepiece.

Solar System The family of planets and other smaller bodies that orbit the Sun. The Solar System currently has eight recognized planets, a handful of dwarf planets, and a huge number of other smaller bodies and moons.

Solar wind The steady stream of charged particles leaving the Sun. Its interaction with the Earth's magnetic field is responsible for the formation of aurorae close to the Earth's poles.

Spectrum The name given to the different radiation wavelengths emitted by a celestial object such as a star or galaxy. An object's spectrum gives clues as to its chemical composition, movement, and even temperature.

Spectral line Emission or absorption of radiation at a distinct wavelength in a spectrum. Spectral lines can be thought of as the fingerprints of different chemicals in an object; they can be used to study the chemical composition of distant stars, galaxies, or planetary atmospheres.

Spiral galaxy A galaxy that has a distinct set of spiral arms composed of bright young stars. Spiral galaxies are rich in gas and dust and offer prime conditions for star formation. Barred spiral galaxies have a central bar-shaped region of stars.

Star A huge sphere of glowing plasma powered by nuclear reactions at its center. Our Sun is a star of medium size. The most massive star known has a mass over 150 times that of the Sun.

Sunspot A region of intense magnetic activity in the Sun's photosphere. It appears darker than the surrounding "surface" because it is much cooler. Sunspots often come in pairs, representing regions of north and south polarity.

Supernova remnant The scattered remnants of a dead massive star. Many supernova remnants are visible with an amateur telescope, including the Crab Nebula in Taurus, the remnant of a supernova that exploded in 1054.

Variable star A star whose magnitude varies over time. The change in brightness may be due to intrinsic changes, or extrinsic ones, such as being eclipsed by another star.

White dwarf The dense, intensely hot glowing star left when a star of similar mass to our Sun dies. As the star dies, it sheds its outermost layers—to make a glowing planetary nebula—and eventually it becomes a white dwarf.

Zenith The point on the sky directly above an observer, and therefore the point on the sky that has an altitude of 90°.

Zodiac The strip of sky through which the Sun appears to pass as it travels on its yearly journey around the sky.

INDEX

Page numbers in **bold** indicate main illustrated references. Page numbers in *italic* indicate other illustrations.

ACKNOWLEDGMENTS

Anton Vamplew
Many thanks to those who have been involved with this book, especially my wife, Gillian, and children, Morten and Etienne. For additional inspiration, my gratitude goes out to: Rob, Monika, David, Tash, Mike, Nina, Katy and Mike W., Keith, Carmel, Trevor, Cherry, Andy, Jane, Emma, Jamie, Tracy, and Tremaine. Finally, a big thanks to Marek and all at Cobalt id for their faith!

Will Gater
I'd like to thank my friends, family, and work colleagues for their support while I've been working on this book; Marek, Louise, and Sarah at Cobalt id for their hard work; and Peter Frances at Dorling Kindersley for his expert guidance.

Publisher's acknowledgments
For their work on the first edition, Dorling Kindersley would like to thank Francis Wong for initial design work; Peter Frances (senior editor); Sarah Larter (managing editor); Michelle Baxter (managing art editor); Mark Cavanagh (jacket designer); Tony Phipps (production editor); and Inderjit Bhullar (senior production controller). Dorling Kindersley would like to thank Surya Sankash Sarangi for picture research for the second edition. For their work on the third edition, Dorling Kindersley would like to thank Nobina Chakravorty and Debjyoti Mukherjee for design assistance; Suefa Lee (Senior Editor); Janashree Singha (Senior Editor); Tarun Sharma (Senior DTP Designer); Priyanka Sharma (Jackets Editorial Coordinator); and Saloni Singh (Managing Jackets Editor).

Cobalt id thank Dorothy Frame for preparing the index and Jacqueline Mitton for editorial guidance; Lee Sproats and Neil Parker at Green Witch, Cambridge, who provided their facility, telescopes, binoculars, and other equipment, as well as their huge expertise during photography. Visit Green Witch at www.green-witch.com.

Thanks also to Michael Hattey at Starlight-xpress, Robert Crawford at Optical Vision Ltd (ww.opticalvision.co.uk), and Cor Berrevoets (www.astronomie.be/registax) for images supplied, and Rachel Pillai for modeling so patiently.

PICTURE CREDITS

Dorling Kindersley would like to thank the following for their permission to reproduce their photographs.

Key: t=top; c=center; b=bottom; l=left; r=right; a=above; B=below.

1c Corbis: Stocktrek Images. **6–7ca** Corbis: Reuters. **8–9** Getty: Stocktrek Images. **10tr** Corbis: Ed Darack/Science Faction; **10c** NASA: G.Schneider, E. Young, G. Rieke, A. Cotera, H.Chen, M.Rieke, R. Thompson; **10–11c** Corbis: NASA/JPL-Caltech. **11br** HST: NASA, ESA, and S. Beckwith (STScI) and the HUDF Team. **12cB** Corbis: Christie's Images; **12br** Corbis: Tony Hallas; **12tr** SPL: David Nunuk; **12cB** SPL: Chris Butler. **12–13c** SPL: Chris Cook. **13br** NASA: NASA/JPL-Caltech. **14tr** Corbis: Roger Ressmeyer; **14br** Corbis: Bettmann. **15bl** Corbis:

Historical Premium. **16tr** Corbis: Fred Hirschmann/Science Faction; **16bc** SPL: John Sanford. **17bl** Corbis: Dennis di Cicco; **17br** Getty: Kauko Helavuo; **17br** HST: Fred Walter (State University of New York at Stony Brook); **17blc** NASA: NASA, ESA, and R. Kirshner (Harvard-Smithsonian Center for Astrophysics). **18tr** HST: NASA, ESA and the Hubble SM4 ERO Team; **18bl** SPL: Mark Garlick. **19bc** HST: F. Paresce, R. Jedrzejewski (STScI) NASA/ESA; **19br** HST: F. Paresce, R. Jedrzejewski (STScI) NASA/ESA; **19tl** SPL: John Chumack. **20cl** Will Gater. **20br** Corbis: Richard Crisp/Science Faction. **20–21c** ESA / Hubble: NASA / The Hubble Heritage Team. **21tc** Corbis: Tony Hallas/ Science Faction; **21tr** Corbis: Stocktrek Images; **21bc** Corbis: Stocktrek Images; **21bl** Getty: Robert Gendler/Visuals Unlimited. **22ca** SPL: Celestial Image Co.; **22cB** SPL: NOAO / AURA / NSF; **23cla** Corbis: Tony Hallas/Science Faction; **23cl** Corbis: Stocktrek Images; **23cr** Corbis: Stocktrek Images; **23bl** Corbis: Stocktrek Images; **23br** Corbis: Stocktrek Images; **23crB** Robert Gendler; **23cra** SPL: Eckhard Slawik; **23clB** SPL: Eckhard Slawik. **24–25c** ESA / Hubble & NASA. **26bl** HST: NASA, ESA and the Hubble SM4 ERO Team; **26blc** HST: NASA, ESA, and the Hubble Heritage Team (STScI/AURA); **26brc** HST: NASA, ESA, and the Hubble Heritage Team (STScI/AURA); **26br** HST: Laurent Drissen, Jean-Rene Roy and Carmelle Robert. **26–27c** Corbis: Encyclopedia. **27bl** Corbis: Roger Ressmeyer; **27cr** HST: John Bahcall, Mike Disney and NASA/ESA. **28cl** SPL: John Chumack; **28cr** SPL: John Chumack; **28c** SPL: Larry Landolfi. **29cla** SPL: NASA / JPL-Caltech / S Stolovy; **29ca** SPL: Eckhard Slawik. **30–31c** Corbis: Frank Lukasseck. **32tr** Corbis: Stapleton Collection; **32b** Getty: Panoramic Images. **33cra** Corbis: Kerrick James; **33cr** Corbis: Amana; **33crB** Getty: Takanori Yamakawa / Sebun Photo; **33cl** istockphoto: Stiv Kahlina. **34tr** istockphoto: Christian Miller; **34cl** istockphoto: Stiv Kahlina. **34–35c** Courtesy of Peter Wienerroither. **35cl** Corbis: Stapleton Collection. **37tr** SPL: Dr Fred Espenak. **38–39c** Corbis: Louie Psihoyos/ Science Faction. **40cl** Corbis: The Gallery Collection; **40tr** istockphoto: Oksana Struk; **41b** Corbis: Alessandro della Bella. **42cr** Archives Charmet: The Bridgeman Art Library; **42tr** Corbis: Tony Hallas/Science Faction; **42bl** SPL: Science Photo Library. **44tr** Corbis: Steve Nagy/ Design Pics; **44cl** Corbis: Tony Hallas/Science Faction; **44br** istockphoto: evirgen. **45bl** Corbis: Dennis di Cicco; **45tc** SPL: Frank Zullo. **46–47c** Getty: Andy Caulfield. **48tr** Corbis: Roger Ressmeyer; **48br** Corbis: Roger Ressmeyer; **48cl** Getty: Thomas Backer. **49cr** Corbis: Stocktrek Images; **49bl** Getty: Jamie Cooper; **49bc** www. perezmedia.net: Jeremy Perez; **49crB** www.perezmedia.net: Jeremy Perez. **50tr** Corbis: Roger Ressmeyer; **50clB** Getty: Robert Postma. **52bl** GPL: Robin Scagell; **52cla** istockphoto: Manfred Konrad; **52tr** istockphoto: Kimeveruss; **52cr** istockphoto: Kristian Sekulic; **52cra** Optical Vision Ltd: Optical Vision Ltd; **52crB** Optical Vision Ltd: Optical Vision Ltd. **54tr** Will Gater. **55tl** Will Gater. **55cl** Will Gater. **56cB** Corbis: Keren Su; **56cr** Corbis: Keren Su; **56bc** Corbis: Keren Su; **56tr** Getty: John Lund; **56cra** Getty: John Lund. **57tc** Corbis: Roger Ressmeyer; **57tr** www.perezmedia.net: Jeremy Perez. **57br** Corbis: Richard Crisp/Science Faction. **60cla** Corbis: Roger Ressmeyer; **60tr** Getty: Keith Douglas; **60ca** SPL: Eckhard Slawik. **62cl** Getty: Handout. **63t** Corbis: HO/Reuters. **64–65c** Alamy: J Marshall - Tribaleye Images. **66tr** SPL: John Chumack; **66bl** Will Gater; **66cr** Swarovski Optik KG. **67br** Corbis: Robert Llewellyn; **67c** www.ilangainc.com: courtesy of iLanga, Inc. **67cla** Will Gater; **68cra** Corbis: Roger Ressmeyer; **68tr** Corbis: Roger Ressmeyer; **68br** Will Gater; **68cr** Will Gater; **68clb** Will Gater; **68bl** Will Gater. **69c** Corbis: Tony Hallas/Science Faction. **70–71c** Getty: Getty Images. **72bl** Corbis: Keren Su; **72tr** Getty: Sisse Brimberg & Cotton Coulson. **73cr** Corbis: Stapleton Collection; **73bl** Wikimedia Commons: U.S. Naval Observatory Library. **74bc** SPL:

NOAO / AURA / NSF. **78–79b** HST: A. Fujii. **80–81c** HST: ESO and Digitized Sky Survey 2. **82tr** Anthony Ayiomamitis; **82br** Getty: LWA. **83bc** Corbis: Roger Ressmeyer; **83br** Corbis: Tony Hallas/Science Faction; **83cB** Getty: LWA. **86cB** Getty: National Geographic/Getty Images. **87ca** Getty: Michael Melford. **88tr** Anthony Ayiomamitis; **88cB** Getty: Stocktrek Images. **89cr** Robert Gendler; **89tl** Robert Gendler; **89br** SPL: Jean-Charles Cuillandre. **90cla** Anthony Ayiomamitis; **90br** GPL: Robin Scagell. **91cr** Matt Ben Daniel. **92cr** Corbis: Roger Ressmeyer; **92tl** GPL: Jeremy Perez. **93cr** Anthony Ayiomamitis; **93br** Corbis: Tony Hallas/Science Faction. **94ca** GPL: Robin Scagell; **94br** GPL: Robin Scagell. **95br** Corbis: Roger Ressmeyer; **95ca** GPL: Robin Scagell. **96c** Corbis: STScI/NASA; **96bl** Corbis: STScI/NASA; **96br** Getty: Stocktrek Images. **97br** GPL: Robin Scagell; **97tl** SPL: NOAO. **98br** SPL: Dr Rudolph Schild. **99cl** GPL: Robin Scagell; **99br** Wikisky: NASA/ESA. **100br** Corbis: Stocktrek Images; **100tr** SPL: NOAO / AURA / NSF. **101cB** Corbis: Richard Crisp/ Science Faction. **102cr** Getty: LWA; **102tr** SPL: Eckhard Slawik. **103tc** Getty: Science & Society Picture Library; **103br** Getty: Stocktrek Images. **104tc** Corbis: Roger Ressmeyer; **104br** SPL: Science & Society Picture Library. **105clB** Anthony Ayiomamitis; **105cr** Corbis: Stocktrek Images; **105br** Getty: Stocktrek Images. **108ca** GPL: Robin Scagell; **108cB** SPL: John Chumack. **109cr** Getty: SSPL. **110br** SPL: John Sanford; **110t** SPL: Jerry Lodriguss. **111bl** Getty: Robert Gendler/ Visuals Unlimited; **111b** Getty: Getty Images; **111cr** SPL: Celestial Image Co. **112cB** GPL: Robin Scagell; **112tr** Robert Gendler. **113br** Corbis: Radius Images. **116tc** Corbis: Roger Ressmeyer; **116bl** GPL: Robin Scagell. **117cB** Anthony Ayiomamitis; **117br** Corbis: Stocktrek Images. **118–119c** Corbis: Tony Hallas/Science Faction. **120ca** GPL: Robin Scagell; **120br** SPL: Robert Gendler. **121br** Anthony Ayiomamitis; **121cra** Corbis: Stapleton Collection. **122bl** Anthony Ayiomamitis; **122cB** Corbis: Stocktrek Images; **122tr** Getty: Robert Gendler/Visuals Unlimited. **123br** Corbis: Stapleton Collection. **124tl** Anthony Ayiomamitis; **124br** Anthony Ayiomamitis; **125crB** Getty: Stocktrek Images; **125** SPL: Mark Garlick. **126bl** Anthony Ayiomamitis; **126tc** Getty: Stocktrek Images. **127bl** Corbis: STScI/NASA; **127cr** SPL: Eckhard Slawik. **128ca** Anthony Ayiomamitis. **129cr** Anthony Ayiomamitis; **129br** Getty: Time & Life Pictures/Getty Images; **129cl** SPL: John Chumack. **130bc** Getty: Robert Gendler/Visuals Unlimited; **130cr** SPL: NOAO / AURA / NSF. **131cr** Corbis: Tony Hallas/Science Faction. **132tr** Corbis: Stocktrek Images; **133cl** Wikimedia Commons: European Southern Observatory. **134tr** SPL: John Chumack; **134b** SPL: John Chumack. **135c** Corbis: Richard Crisp/Science Faction; **135bc** Corbis: NASA - Hubble Heritage Team; **135br** Corbis: Roger Ressmeyer. **138cra** Corbis: TAKASHI KATAHIRA/amanaimages; **138cB** SPL: Eckhard Slawik. **139cl** GPL: Robin Scagell; **139bl** Getty: Stocktrek Images; **139cr** SPL: Celestial Image Co. **140–141cl** GPL: Robin Scagell. **142br** Corbis: STScI/NASA; **142cB** GPL: Robin Scagell; **142bc** Getty: Stocktrek Images. **142–143c** GPL: Robin Scagell. **143tl** Corbis: Stocktrek Images. **144cl** SPL: NOAO. **145br** NASA: NASA, ESA, N. Smith (Univ. of California, Berkeley), Hubble Heritage Team (STScI/AURA); **145ca** Corbis: Celestial Image Co.; **145tl** SPL: Celestial Image Co. **146–147c** Corbis: Stocktrek Images. **148br** Corbis: STScI/NASA; **148cB** HST: Hubble Heritage Team (STScI/AURA/NASA/ESA); **148tr** Robert Gendler. **149bc** Corbis: Stocktrek Images; **149cB** Getty: Stocktrek Images; **149cr** NASA: AAO: Photograph by David Malin. **150cr** Corbis: Bettmann. **151cB** Corbis: Dennis di Cicco; **151bl** Getty: Robert Gendler/Visuals Unlimited. **152cr** SPL: Royal Observatory, Edinburgh. **153bc** Corbis: Stocktrek Images; **153tc** Getty: Robert Gendler/Visuals Unlimited; **153br** HST: Ron Gilliland (Space Telescope Science Institute) and NASA/ESA. **154cr** Art Archive: Museo del Prado Madrid. **155br** Getty: Robert Gendler/Visuals Unlimited; **155bl** NASA: NASA/STScI/WikiSky. **156cra** GPL: Robin Scagell. **157cB** Anthony Ayiomamitis. **158tc** Anthony Ayiomamitis; **158cr** SPL: NOAO / AURA / NSF. **159tr** GPL: Robin Scagell; **159bl** SPL: NOAO. **160bc** Getty: Stocktrek Images; **160br** HST: NASA, ESA and G. Bacon (STScI). **161crB** GPL: Robin Scagell; **161c** NASA: NASA/JPL-Caltech/T. Currie (CfA). **162tc** HST: Margarita Karovska (Harvard-Smithsonian Center for Astrophysics), and NASA/ESA; **162bc** NASA. **163bl** Getty: Stocktrek Images. **164c** Credner: Till Credner. **164–165cB** NASA. **165cl** GPL: Robin Scagell; **165cl** NASA; **165cr** SPL: NOAO / AURA / NSF. **166br** Anthony Ayiomamitis. **167c** Corbis: Stocktrek Images; **167cr** NASA. **168br** Getty: Robert Gendler/Visuals Unlimited; **168tr** SPL: Jerry Lodriguss. **170cB** GPL: Robin Scagell. **171br** Robert Gendler. **172–173c** Getty: Getty/Digital Vision. **174br** Getty: AFP/Getty Images. **174–175c** SPL: Larry Landolfi. **176cr** Corbis: NASA/Corbis; **176b** SPL: The International Astronomical Union. **176–177tl** GPL: Robin Scagell. **177bl** Corbis: NASA/Corbis; **177tc** GPL: Robin Scagell; **177cla** GPL: Robin Scagell; **177ca** GPL: Robin Scagell. **178cr** GPL: Robin Scagell; **178tl** SPL: Eckhard Slawik; **178ca** Wikimedia Commons: NASA/JPL. **179br** Corbis: Bettmann; **179tr** Corbis: NASA/JPL-Caltech; **179tr** Corbis: NASA/Corbis; **179c** Getty: Digital Vision. **180bl** Corbis: NASA/ JPL-Caltech; **180bl** Corbis: NASA/Corbis; **180tr** Corbis: William Radcliffe/Science Faction; **180tr** NASA. **180–181cB** Corbis: Bettmann. **181br** Corbis: Michael Benson/Kinetikon Pictures; **181c** Corbis: William Radcliffe/Science Faction. **182br** GPL: Robin Scagell; **182br** HST: NASA/ESA, The Hubble Heritage Team; **182tr** NASA: JPL. **182–183tr** Corbis: NASA - Hubble Space Telescope - /Science Faction. **183c** Corbis: Bettmann. **184–185** NASA: JPL-Caltech / SwRI / MSSS / Betsy Asher Hall / Gervasio Robles. **186bl** Corbis: NASA/Corbis; **186tr** Corbis: STScI/NASA; **186tr** Getty: Stocktrek Images. **186–187cr** SPL: John Chumack. **187c** Corbis: NASA - digital version copyright/Science Faction; **187br** NASA: JPL/STScI. **188cr** Corbis: Bettman; **188bl** Corbis: NASA/Corbis; **188br** Getty: StockTrek. **189tr** Corbis: NASA/Corbis; **189tr** SPL: NASA; **189tc** W.M. KECK Observatory. **190tr** JHUAPL / SWRI; **190cla** Dorling Kindersley: PAL; **190cB** Dorling Kindersley: PAL; **190bl** JPL-CALTECH / UCLA / MPS / DLR / IDA. **191l** Corbis: EPA. **192cB** PAL: Dorling Kindersley; **192bc** PAL: Dorling Kindersley; **192br** PAL: Dorling Kindersley; **192tl** SPL: John Chumack. **193cra** Corbis: George Shelley; **193cB** Corbis: Stocktrek. **194–195c** Corbis: Roger Ressmeyer. **196c** Getty: Chris Ware; **196cl** projectfullmoon.com: Michael Light; **196clB** SPL: John Sanford; **196–197tl** Corbis: Jeff Vanuga. **197c** Corbis: Stocktrek Images; **197bc** PAL: Dorling Kindersley; **197br** PAL: Dorling Kindersley. **198tr** Alamy: J Marshall - Tribaleye Images. **199cla** Anthony Ayiomamitis; **199crB** Anthony Ayiomamitis; **199tr** Corbis: Bettmann; **199bl** George Tarsoudis; **199clB** George Tarsoudis; **199cra** George Tarsoudis; **199br** George Tarsoudis. **200b** Corbis: Aaron Horowitz; **200cl** NASA: JPL/Brown Univ. **201cl** Corbis: Tony Hallas/Science Faction; **201tr** Corbis: Tony Hallas/Science Faction; **201br** PAL: Dorling Kindersley. **202c** Corbis: Hinrich Baesemann/dpa; **202cl** SPL: Stephen J. Krasemann; **202br** SPL: Pekka Parviainen. **203tr** Corbis: Comstock Select. **204c** SPL: NASA. **204–205tl** SPL: David Ducros. **205br** SPL: Phil Dauber. **206–207c** HST: NASA, ESA, and H. Bond (STScI). **208tr** SPL: Science Source. **209c** Corbis: Aaron Horowitz. **234–235r** Corbis: Tony Hallas/Science Faction. **237b** SPL: NOAO / AURA / NSF. **238bl** SPL: Jerry Schad. **243br** Corbis: Tony Hallas/Science Faction.

All other images © Dorling Kindersley Ltd. For further information, see www.dkimages.com